A QUESTION OF BALANCE

Private Rights and the Public Interest in Scientific and Technical Databases

Committee for a Study on Promoting Access to
Scientific and Technical Data for the Public Interest

Commission on Physical Sciences,
  Mathematics, and Applications

National Research Council

NATIONAL ACADEMY PRESS
Washington, D.C.

Support for this project was provided by the National Science Foundation (under grant no. OCE-9729508), the National Institutes of Health (under purchase order no. 467-MZ-801699), the National Institute of Standards and Technology (under contract no. 43NANB909028), the National Aeronautics and Space Administration, the National Oceanic and Atmospheric Administration, and the United States Geological Survey (through the aforementioned National Science Foundation grant no. OCE-9729508), and the Department of Energy (under contract no. DE-FG02-96ER30277).

International Standard Book Number 0-309-06825-8

Library of Congress Catalog Card Number 99-68421

Additional copies of this report are available from National Academy Press, 2101 Constitution Avenue, N.W., Lockbox 285, Washington, D.C. 20055; (800) 624-6242 or (202) 334-3313 (in the Washington metropolitan area); Internet <http://www.nap.edu>.

THE NATIONAL ACADEMIES

National Academy of Sciences
National Academy of Engineering
Institute of Medicine
National Research Council

The **National Academy of Sciences** is a private, nonprofit, self-perpetuating society of distinguished scholars engaged in scientific and engineering research, dedicated to the furtherance of science and technology and to their use for the general welfare. Upon the authority of the charter granted to it by the Congress in 1863, the Academy has a mandate that requires it to advise the federal government on scientific and technical matters. Dr. Bruce M. Alberts is president of the National Academy of Sciences.

The **National Academy of Engineering** was established in 1964, under the charter of the National Academy of Sciences, as a parallel organization of outstanding engineers. It is autonomous in its administration and in the selection of its members, sharing with the National Academy of Sciences the responsibility for advising the federal government. The National Academy of Engineering also sponsors engineering programs aimed at meeting national needs, encourages education and research, and recognizes the superior achievements of engineers. Dr. William A. Wulf is president of the National Academy of Engineering.

The **Institute of Medicine** was established in 1970 by the National Academy of Sciences to secure the services of eminent members of appropriate professions in the examination of policy matters pertaining to the health of the public. The Institute acts under the responsibility given to the National Academy of Sciences by its congressional charter to be an adviser to the federal government and, upon its own initiative, to identify issues of medical care, research, and education. Dr. Kenneth I. Shine is president of the Institute of Medicine.

The **National Research Council** was organized by the National Academy of Sciences in 1916 to associate the broad community of science and technology with the Academy's purposes of furthering knowledge and advising the federal government. Functioning in accordance with general policies determined by the Academy, the Council has become the principal operating agency of both the National Academy of Sciences and the National Academy of Engineering in providing services to the government, the public, and the scientific and engineering communities. The Council is administered jointly by both Academies and the Institute of Medicine. Dr. Bruce M. Alberts and Dr. William A. Wulf are chairman and vice chairman, respectively, of the National Research Council.

Preface

In response to a request from several federal science agencies, the Committee for a Study on Promoting Access to Scientific and Technical Data for the Public Interest (see Appendix A) undertook a study to identify and evaluate the various existing and proposed policy approaches (including related legal, economic, and technical considerations) for protecting the proprietary rights of private-sector database rights holders while promoting and enhancing access to scientific and technical (S&T) data for public-interest uses. Specifically, the sponsors asked the study committee to address the following issues:

1. Describe the salient characteristics and importance of scientific and technical databases in research, both in general categories and using specific examples.

2. Describe the practices of the production, dissemination, and use of S&T data in the federal, nonprofit, and commercial contexts.

3. Identify the major incentives and disincentives in the production, dissemination, and use of S&T data in the federal, not-for-profit, and commercial contexts.

4. Review the key elements of existing and proposed intellectual property rights regimes for noncopyrightable databases and other "collections of information," including technical protection measures, with specific emphasis on S&T databases. Also review the federal government policies regarding scientific data production, protection, dissemination, and use, particularly for data produced or disseminated by nongovernment entities under an agreement with government, including with government funding.

5. Consider the pros and cons of legal, policy, and technical options identi-

fied in response to item 4 above, with particular attention to balancing the interests of S&T database providers and disseminators in protecting their investments with the interests of promoting access to and use of S&T data for research and other public-interest uses.

6. Identify issues that require further analysis and resolution, and how to address them.

7. Provide conclusions and recommendations where possible, or otherwise provide an assessment of options.

In discussing the charge and the scope of the project with the committee, the sponsors asked the committee to focus in particular on the legislative proposals on database protection then pending in Congress as examples of the kinds of statutory options that might be adopted. Both the sponsors and the committee were well aware of the fact that those pending proposals would change further and therefore presented "moving targets" for the study. It is for this reason that the committee's recommendations regarding any potential legislation in this area are offered as guiding principles rather than as specific language for a specific bill.

The focus of the study was further constrained to domestic, rather than international, issues. The committee was cognizant of the fact that any new U.S. legislation would ultimately have substantial significance internationally, both in the economic and legal domains and in the S&T research community, but it limited its investigation and analysis of foreign laws and international legal issues, concentrating only on their direct influence on the U.S. domestic legal and policy situation. In addition, although the subject matter included all S&T databases, the committee was able to choose only representative examples for discussion and analysis in the report. For instance, the committee did not include specific examples from the social sciences or the space sciences, among other disciplines. Nevertheless, the committee believes that the relatively broad spectrum of S&T databases that it did use captured the most significant issues in the context of database protection and public-interest uses.

In responding to its charge, the committee made significant efforts to obtain broad input from representatives of the main identified interest groups, primarily through a workshop that was held on January 14-15, 1999, at the main Department of Commerce building in Washington, D.C. (Appendix B gives the agenda and lists the participants). The workshop *Proceedings*—taped, transcribed, edited, and published only in electronic form on the National Academies' Web site at <www.nap.edu>—were a major source of information for this report (Appendix C lists the contents of the published *Proceedings*).[1] The committee also met

---

[1] The views expressed in the committee's workshop *Proceedings* are solely those of the individual authors and workshop participants. The separate *Proceedings* report does not provide conclusions and recommendations.

on two other occasions to gather information and to work on this report. In addition, the underlying technical factors and developments associated with digital networked information, and their impact on intellectual property rights protection, are examined in detail in a concurrent study, *The Digital Dilemma: Intellectual Property in the Information Age* (National Academy Press, Washington, D.C., 2000, in press), by the National Research Council's Computer Science and Telecommunications Board.

The report that follows reflects the deliberative consensus of the study committee. It is our hope that the committee's conclusions and recommendations will help the sponsors of the study, the legislators examining database protection proposals, and the broader S&T community to understand better the issues in striking a proper balance between protecting rights in and promoting public-interest uses of scientific and technical databases.

Robert J. Serafin, *Chair*
Paul F. Uhlir, *Study Director*

Acknowledgments

The study committee wishes to express its sincere thanks to the many individuals who played significant roles in the completion of this study. The committee sponsored the Workshop on Promoting Access to Scientific and Technical Data for the Public Interest: An Assessment of Policy Options on January 14-15, 1999, at the Department of Commerce in Washington, D.C., and it extends its thanks to the following individuals who made presentations during the January 14, 1999, plenary session: Q. Todd Dickinson, acting commissioner of Patents and Trademarks, Department of Commerce, gave the keynote address; Barbara Ryan of the U.S. Geological Survey and Barry Glick, formerly with GeoSystems Global Corporation, participated in the geographic data panel; G. Christian Overton of the University of Pennsylvania's Center for Bioinformatics, James Ostell of the National Library of Medicine's National Center for Biotechnology Information, and Myra Williams of the Molecular Applications Group participated in the genomic data panel; Richard Kayser of the National Institute of Standards and Technology, James Lohr of the American Chemical Society's Chemical Abstracts Service, and Leslie Singer of the Institute for Scientific Information, Inc. participated in the panel that discussed chemical and chemical engineering data; Kenneth Hadeen of the National Oceanic and Atmospheric Administration's National Climatic Data Center (retired), David Fulker of the University Corporation for Atmospheric Research's Unidata Program, and Robert Brammer of TASC participated in the meteorological data panel; Richard Gilbert of the University of California at Berkeley discussed economic factors in the production, dissemination, and use of scientific and technical (S&T) databases in the public and private sectors; Stephen Maurer, attorney, submitted a commissioned paper for the study (reproduced as Appendix C of the online workshop *Proceedings*); Teresa Lunt of the Xerox Palo Alto Research Center provided an overview of the current situation and future prospects with respect to technologies for protecting and also for misappropriating digital intellectual property rights; Marybeth Peters, Register of Copyrights in the Library of Congress, provided a summary overview of the existing and proposed intellectual property rights regimes for databases; and Justin Hughes, of the Department of Commerce's Patent and Trademark Office, summarized the relevant federal government information law and data policies. The aforementioned data panelists also participated in the January 14, 1999, discussion sessions on not-for-profit-sector data, government-sector data, and commercial-sector data.

The committee would also like to thank those who participated as panelists in the January 15, 1999, discussion sessions on the potential impacts of legislation and assessments of policy options during the workshop. Jon Baumgarten of Proskauer Rose, LLP, Peter Jaszi of the American University School of Law, James Neal of the John Hopkins University Library, and Ferris Webster of the University of Delaware joined Kenneth Hadeen, David Fulker, and Robert Brammer in discussing what would happen should Congress decide to enact a strong property rights model for protecting databases. Dennis Benson of the National Center for Biotechnology Information, Jonathan Band of Morrison & Foerster, LLP, and Thomas Rindfleisch of Stanford University's Lane Medical Library discussed, with Chris Overton and Myra Williams, the possible scenarios should Congress enact an unfair competition model for protecting databases. Prue Adler of the Association of Research Libraries, Eric Massant of Reed Elsevier, Inc., Tim Foresman of the University of Maryland, and Kenneth Frazier of the University of Wisconsin Libraries joined Barry Glick in a discussion assessing legal and policy options in promoting access to and use of government S&T data for the public interest. Finally, Jerome Reichman of the Vanderbilt University School of Law and R. Stephen Berry of the University of Chicago discussed possible legal and policy options associated with promoting access to and use of not-for-profit-sector S&T data for the public interest with Richard Kayser, James Lohr, and Leslie Singer. The committee is also very appreciative of the contributions of more than 100 individuals who attended the workshop. In addition, it extends its gratitude to Jean Schiro-Zavela of the National Oceanic and Atmospheric Administration and to Justin Hughes of the Patent and Trademark Office for helping to make arrangements for the workshop.

The committee would also like to express its gratitude to the following ex officio members, who provided liaison with other relevant activities: Goetz Oertel of the Association of Universities for Research in Astronomy and chair of the U.S. National Committee for CODATA; Shelton Alexander, of Pennsylvania State University, who is a member of the National Research Council's Computer Science and Telecommunications Board's committee studying intellectual property rights in the networked environment; and Francis Bretherton of the University of Wisconsin, chair of the National Research Council's Committee on Geophysical and Environmental Data.

This report has been reviewed in draft form by individuals chosen for their diverse perspectives and technical and legal expertise, in accordance with procedures approved by the National Research Council's Report Review Committee. The purpose of this independent review is to provide candid and critical comments that will assist the institution in making the published report as sound as possible and to ensure that the report meets institutional standards for objectivity, evidence, and responsiveness to the study charge. The review comments and draft manuscript remain confidential to protect the integrity of the deliberative process. The study committee would like to thank the following individuals for their participation in the review of this report: George Annas of Boston University, Lois Blaine of the American Type Culture Collection, Kenneth Dam of the University of Chicago School of Law, John Estes of the University of California at Santa Barbara, Richard Hallgren of the American Meteorological Society, Michael Keller of Stanford University Library and Highwire Press Inc., Gary King of Harvard University, Charles McClure of Syracuse University, Roger Noll of Stanford University, Pamela Samuelson of the University of California at Berkeley, William Sprigg of the University of Arizona, Hal Varian of the University of California at Berkeley, and Ronald Wigington of the American Chemical Society (retired).

In addition, the following individuals reviewed the workshop *Proceedings:* Bonnie Carroll of Information International Associates, Inc., David Lide, Jr., publishing consultant, and Goetz Oertel of the Association of Universities for Research in Astronomy.

Finally, the committee would like to recognize the contributions of the National Research Council staff without whom this report could not have been completed: Paul Uhlir, director of International Scientific and Technical Information Programs of the Office of International Affairs, who served as study director and organized the workshop and other study committee meetings; Julie Esanu, who provided research and program assistance to the committee, as well as editorial work on the workshop *Proceedings*; Barbara Wright and Pamela Gamble for the staff support to the committee; and Susan Maurizi and Janet Overton, who edited the final committee report and the *Proceedings*.

Contents

A QUESTION
OF BALANCE

Summary

Legislative efforts are currently under way in the United States, the European Union, and the World Intellectual Property Organization (WIPO) to greatly enhance the legal protection of proprietary databases. One of these efforts, the European Union's 1996 Directive on the Legal Protection of Databases,[1] already has been finalized and is now being implemented by the European Union's member states, while other legislative initiatives in the U.S. Congress and WIPO are still pending action. As discussed in detail in the 1997 report *Bits of Power: Issues in Global Access to Scientific Data*[2] and in subsequent publications,[3] these new legal approaches threaten to compromise traditional and customary access to and use of scientific and technical (S&T) data for public-interest endeavors, including not-for-profit research, education, and general library uses. At the same time, there are legitimate concerns by the rights holders in databases regarding unauthorized and uncompensated uses of their data products, including at times the wholesale commercial misappropriation of proprietary databases.

---

[1] The E.U. Database Directive—Directive 96/9/EC of the European Parliament and of the Council of 11 March 1996 on the Legal Protection of Databases, 1996 O.J. (L77) 20. The E.U. Database Directive is reprinted as Appendix D of this report.

[2] National Research Council (1997), *Bits of Power: Issues in Global Access to Scientific Data*, National Academy Press, Washington, D.C.

[3] See, e.g., J.H. Reichman and Paul F. Uhlir (1999), "Database Protection at the Crossroads: Recent Developments and Their Impact on Science and Technology," *Berkeley Technology Law Journal*, Vol. 14, No. 2, p. 793, and Stephen M. Maurer and Suzanne Scotchmer, "Database Protection: Is It Broken and Should We Fix It?" *Science*, Vol. 284, p. 1129.

Factual data are both an essential resource for and a valuable output from scientific research. It is through the formation, communication, and use of facts and ideas that scientists conduct research and educate students. Our nation has a vibrant and demonstrably productive sector of S&T database producers, disseminators, and users that has led the world. Advances in computing and communications technologies make S&T databases and the facts they contain increasingly valuable for producing new discoveries and for accelerating the growth of knowledge and the pace of innovation. The same technologies that facilitate the effective production, dissemination, and use of digital databases, however, can also expedite their unauthorized dissemination and use, undermining incentives to create new databases, enabling unfair competition and wholesale misappropriation, and in the most extreme cases, exposing the original database rights holder to market failure.

The institutions, organizations, and individuals involved form a highly complex web of interrelationships—some competing and some complementary, some creating or using original data collections or derivative productions, and spanning activities in both public and private, national and international, and scientific and non-scientific contexts (see Table S.1 for some representative examples of different types of S&T database activities). Although these diverse actors all have their own goals and motivations, they nonetheless may be broadly characterized in three fairly distinct groups. The first is the government sector, which produces S&T databases as a public good and has a mandate under OMB Circular A-130[4] to disseminate the fruits of those activities as broadly and openly as possible, and to provide efficient access. The second is the commercial sector, which produces databases as a private good and typically maintains those data on a proprietary and restricted basis, either for internal purposes or for commercial vending, with the goal of full cost recovery plus profit. The third is the not-for-profit sector, which includes universities, research institutes, and various public-interest organizations that produce databases in support of their institutional mission and typically disseminate data on a cost-recovery basis, which can cause them to take a middle ground in terms of treating their databases as a public or a private good.

As users of databases, all three sectors support the public-good approach of the federal government to data distribution, and all seek to minimize the costs they may need to pay for access to and use of data from private-sector sources. **A principal concern of the committee is that the development of any new database protection measures directed toward protecting private-sector investments take into account the need to promote access to and subsequent use of S&T data and databases not only by the not-for-profit sector, but by for-profit creators of derivative databases as well.**

---

[4] Office of Management and Budget (1993), Circular A-130, "Management of Federal Information Resources," U.S. Government Printing Office, Washington, D.C.

Numerous legal, technical, and market-based approaches already exist to protect proprietary rights in databases. Existing legal measures include (1) copyright law, recently updated and strengthened for the online digital environment; (2) licensing, a subset of contract law, which is increasingly the method of choice for online vendors of proprietary databases and other information products; (3) trade secret law, used in conjunction with contract law and various new technological protections; and (4) unfair competition law in state common law, which is of limited value for database protection at this time but is viewed as a potential model for a new federal database protection statute. These legal measures are being supported with increasing efficiency online by technological protections such as digital encryption, watermarks, download limitations, access controls, and both hardware- and software-based trusted systems. Finally, there are important market-based approaches to protecting databases, such as frequent updating or customizing of database content, that can help prevent stolen databases from undermining the rights holder's market for very long or very broadly.

For public-sector databases, including S&T databases, there are well-established laws and policies in place[5] that generally prohibit proprietary protections of databases by the government and that treat those data as public goods and promote their full and open availability to the public. While many not-for-profit S&T database producers and vendors—especially those that receive government funding for their activities—adhere to the policy of full and open data availability for public-sector databases, other not-for-profits seek the full protection of the law for their proprietary databases. This division of interests has been further exacerbated by the enactment of the 1996 E.U. Database Directive, which provides strong and unprecedented property rights in public- and private-sector databases and the substantial components of such databases.

The current efforts to enact statutory federal database protection in the United States appear to be stimulated by three principal factors: (1) the possibility for rapid and complete database copying with the potential for instantaneous broad dissemination; (2) a gap in U.S. law created by the *Feist*[6] decision, which served as the final blow to invalidating copyright protection on the basis of "sweat-of-the-brow" investments alone; and (3) the E.U. Database Directive, which requires non-E.U. nations to pass a similar law in order for their citizens to enjoy the directive's protections in Europe, thereby providing a potentially unfair advantage to European competitors of the U.S. private sector.

The committee believes, however, that the need for additional statutory pro-

---

[5] See, e.g., 44 U.S.C., section 3506(d)(1)(A) (Supp. 1995), and Office of Management and Budget (1993), Circular A-130, note 4.

[6] *Feist Publications, Inc. v. Rural Telephone Service Co.*, 499 U.S. 340 (1991). It is important to note, however, that *Feist* did not "overturn" the "sweat-of-the-brow" doctrine under copyright, which Congress had actually done already under the Copyright Act of 1976. Moreover, the sweat-of-the-brow doctrine under state law was never a prevailing legal approach.

TABLE S.1   Examples of Different Types of S&T Database Activities
Discussed in the January 1999 Workshop

Organization (Sector)	Information and Tools Provided	Data Sources
Geographic and Environmental		
U.S. Geological Survey (USGS) (Government)	Geographic data: maps and map products Data from other programs: biologic, geologic, hydrologic	USGS, other federal agencies, state and local governments, not-for-profit researchers, partnerships with private sector
Long-Term Ecological Research (LTER) Network Office (Not-for-Profit)	Site description database, integrated climate database, remotely sensed ecological data	Ecological researchers at distributed sites belonging to the LTER network
GeoSystems Global Corp. (Commercial)	Digital maps, MapQuest Web site, mapping services	From the public domain: government-produced maps (federal, state, local), digital geographic data, remotely sensed imagery Other sources: commercial and other countries' maps, digital data, remotely sensed imagery, other published sources
Genomic		
National Center for Biotechnology Information (Government)	GenBank: DNA and protein sequence data; Other genomic mapping databases; 3D protein structure database; bibliographic databases; software tools	Direct contributions from scientists; access to other databases from government, not-for-profit, other country sources
Center for Bioinformatics University of Pennsylvania (Not-for-Profit)	Specialized biological databases; software tools for integration of distributed heterogeneous databases	Proprietary and public-domain experimental data from academic researchers; manual processing and encoding of data from published literature; online molecular and cellular biology and genomic databases
Molecular Applications Group (Commercial)	Software for storing, mining, and visualizing genomic data; databases derived from public and private data and proprietary software	>150 online database sites, public and proprietary

Users	Dissemination Modes
USGS, other government agencies, commercial database providers and value adders, researchers, and the general public	Maps: hard copy (paper, plastic, film) and digital form; distributed by agency directly and through partnerships with private sector, not-for-profit sector
Researchers	Internet, some tape and CD-ROM for portability
Commercial clients: large companies and consumers	Maps: hard copy and digital form Software products distributed via retail channels (CD-ROM) and directly to corporate customers Mapping services distributed via Internet
Research scientists in academic, government, commercial organizations	Internet access via Web servers and File Transfer Protocol (FTP)
Research scientists in academic, government, commercial organizations (U.S. and abroad)	Internet access Source code distributed directly
Research scientists in academic, government, commercial organizations (U.S. and abroad)	Some software products downloaded from the Web; others require on-site expert installation

continued

Continued

Organization (Sector)	Information and Tools Provided	Data Sources
Chemical and Chemical Engineering		
National Institute of Standards and Technology (NIST) Physical and Chemical Properties Division (Government)	Specialized chemistry and chemical engineering databases (extensively evaluated and documented)	Experimental results from published literature; experiments done specifically for data acquisition; published data evaluations; supplementary data deposits
Chemical Abstracts Service American Chemical Society (Not-for-Profit)	Chemical Abstracts: bibliographic database Registry: registry of chemical substances Software access tools	Journals, patents, books, proceedings, dissertations
Institute for Scientific Information (Commercial)	Bibliographic databases: citation indexes, tables of contents Information services Linkages to publishers' full-text databases	Journals, books, proceedings (print and electronic format)
Meteorological		
National Climatic Data Center (Government)	Climatological summaries from National Weather Service stations; historical long-term climatic databases	National Weather Service, World Meteorological Organization, NASA, bilateral agreements with other countries
Unidata Program, University Corporation for Atmospheric Research (Not-for-Profit)	Quasi-real-time atmospheric and related data Case study data sets Software tools	Public: National Weather Service, National Environmental Data Service Private: network of lightning sensors, sensors in commercial aircraft
TASC (Commercial)	Real-time weather information	Public: National Weather Service—downlink directly from U.S. and international weather satellites, other observational sources

NOTE: Although the subject matter of this study included all S&T databases, the committee was able to choose only representative examples for discussion and analysis in the report. For instance, specific examples from the social sciences or the space sciences, among other disciplines, were not included.

Users	Dissemination Modes
Researchers in academic, government, commercial organizations (some databases used primarily by industrial users)	Variety of forms: hard-copy publication, CD-ROM or floppy disk, Internet access; NIST distributes directly or via agreements with secondary distributors
Researchers in academic, government, commercial organizations; patent examiners; students	Electronic access, hard copy, CD-ROM
Academic, government lab, and corporate libraries; researchers in academic, government, commercial organizations	Diskette, CD-ROM, FTP files, Internet access, hard copy
Individuals, commercial clients, government agencies, engineering uses	Hard-copy, microfiche, magnetic tape, disks, CD-ROM, FTP, Internet
Academic departments	Internet
News media (broadcast and cable TV), aviation, energy and power, agribusiness	Public and private data communication networks: satellite broadcasting services and Internet

tection has not been sufficiently substantiated. The high level of activity in the production and use of digital S&T (and other) databases in the United States serves as prima facie evidence that the threats of misappropriation do not constitute a crisis. Nor do the existing legal, technical, and market-based measures provide a chronic state of underprotection for proprietary databases. The almost universal use of licensing, rather than sale, of online databases and other digital information, coupled with technological enforcement measures, on balance potentially provides much stronger protections to the vendors vis-à-vis their customers than they enjoyed prior to *Feist* and under the print media copyright regime. While some of the current law providing protection to database rights holders remains uncertain in terms of scope of applicability, the trend in recent years has been to broaden, rather than narrow, applicable intellectual property protections. Moreover, for the many reasons discussed in this report, strong statutory protection of databases would have significant negative impacts on access to and use of S&T databases for not-for-profit research and other public-interest uses (see Table S.2). Nevertheless, although the committee opposes the creation of any strong new rights in compilations of factual information, it recognizes that limited new federal legal protection against wholesale misappropriation of databases may be appropriate. In particular, a balanced alternative to the E.U. Database Directive might be achieved in a properly scoped and focused new U.S. law, one that could then serve as a model for an international treaty in this area.

RECOMMENDATIONS

Because of the complex, interdependent relationships among public-sector and private-sector database producers, disseminators, and users, any action to increase the rights of persons in one category likely will compromise the rights of the persons in the other categories, with far-reaching and potentially negative consequences.[7] Of course, it is in the common interest of both database rights holders and users—and of society generally—to achieve a workable balance among the respective interests so that all legitimate rights remain reasonably

---

[7] The potential for unintended consequences from new *sui generis* intellectual property legislation and the need for caution were emphasized by the former chairman of the House Committee on the Judiciary, Robert W. Kastenmeier, in Kastenmeier and Remington (1985), "The Semiconductor Chip Protection Act of 1984: A Swamp or Firm Ground?" *Minn. L. Rev.,* Vol. 70, pp. 417, 440-442, where the authors suggested that any proposal for such legislation should meet a four-point test: (1) that the protection would fit harmoniously with other intellectual property regimes; (2) that the new protection can be defined in a reasonably clear and satisfactory manner; (3) that the new proposal is based on an honest analysis of all the costs and benefits; and (4) that the legislation should show clearly that it would enrich and enhance the "aggregate public domain."

TABLE S.2   Summary Comparison of Not-for-Profit Research User Rights
Under Traditional Copyright and Under Online Licensing When Combined
with *Sui Generis* Database Protection Legislation

Traditional Copyright Law	Licensing Plus *Sui Generis* Protection Such as Provided by the E.U. Database Directive
1. User can immediately use all disparate factual data and information disclosed in a database; copyright law does not protect ideas or facts. Fair-use exception available for certain additional research or educational uses, even of protected expression.	1. Even after paying for access to factual data and information, which are not copyrightable by definition, user faces limitations on use in any ways prohibited by the license and as reinforced by the legislation; user cannot distribute another database, using the same factual data or information, without either seeking permission and perhaps paying another fee or regenerating those protected data independently.
2. User can independently create another version of a database and sell it; copyright law allows independent creation, and all factual data are in the public domain.	2. User can independently create another version of the database. If this is not possible, user needs a license or permission to combine legitimately accessed factual data or information into a derivative data product; the licensor can claim that the user is violating redistribution and other rights, and the user must guess what courts will consider to be a quantitatively or qualitatively insubstantial part of the database; the licensor is under no duty to grant such a license; and if a sole source, the licensor may not want any competition from follow-on products.
3. User can combine noncopyrightable factual data with other data and information into a multiple-source or interdisciplinary database for research or educational purposes without permission or additional payment to the originators.[a]	3. User cannot lend, give, or sell data to others even after paying for access (unless permitted by the license) because there is no first sale, only a license; user would have to obtain express permission and perhaps pay additional fees to avoid the risk of harming the market (e.g., possibly causing one lost sale).
4. User can make limited or "fair use" of even protected expression for not-for-profit research or classroom purposes; such uses often deemed fair or privileged uses under statutory law or precedents.	4. Because there are no limits on licensing, user is subject to database vendor overriding even those exceptions contained in the legislation, including exceptions for research, education, or other public-interest uses.
5. Following the first purchase of a copyrightable database in hard copy, user can lend, give, or resell it to anyone else under the first-sale doctrine, borrow it from a library, use it at any time for virtually any [lawful] purpose, and make a copy of it for personal or scholarly purposes.	5. During the period of protection, user rights depend on the terms of the license supported by the new property right; database would not enter the public domain for at least 15 years (and in Europe possibly never if the rights holder continues to invest in maintenance or updates of a dynamic database).

NOTE: This summary table was compiled from a more detailed comparative discussion presented in an article by J.H. Reichman and Paul F. Uhlir (1999), "Database Protection at the Crossroads: Recent Developments and Their Impact on Science and Technology," *Berkeley Technology Law Journal*, Vol. 14, No. 2, pp. 799-821.
[a] Acknowledgment of sources is an appropriate academic norm, but their express permission is not required.

protected. Therefore, **as a general guiding principle, the committee recommends that any new federal protection of databases should balance the costs and benefits of the proposed changes for both database rights holders and users.** In addition to this general principle, the committee makes a number of recommendations—based on its assessment of legislative options[8] and related policy options—to the government, and it makes one recommendation to the not-for-profit S&T community.

Recommended Legislative Principles

The committee recommends that any new federal statutory protection of databases incorporate the following principles:

1. **Limit any additional protection to prohibition of acts of unauthorized taking that cause substantial competitive injury to the database rights holder in the rights holder's actual market.** The standard of harm should be sufficiently clear to permit good-faith users to know when they are infringing on a database rights holder's rights and should not undermine the nation's capabilities for innovation or competition in the marketplace.

2. **Constrain the subject-matter scope to databases comprising a collection of discrete facts and items of information, and expressly exclude collections of copyrightable material, which is already protected.** Protection under any new statute should extend only to a database that is the product of a substantial investment and not to any idea, fact, procedure, system, method of operation, concept, principle, or discovery disclosed by the database.

3. **Limit the term of protection to a period of time sufficient to provide incentives found necessary for the creation of new databases.** If legislation with a fixed term of protection is adopted, an appropriate term of protection most likely should be substantially shorter than the proposed 15-year term. It also should be based on an analysis of the economics of the database industry, rather than set arbitrarily.

---

[8] In response to its charge, the committee selected the three major legislative models that were introduced into the *Congressional Record* by Senator Orrin Hatch shortly after the committee's January 1999 workshop. See *Cong. Rec.,* Vol. 106, S. 316 (Jan. 19, 1999). The committee evaluated and compared the major provisions of these three models in arriving at the consensus principles and recommendations presented here. The rationale for each recommended principle or action is contained in Chapter 4. In evaluating the three legislative proposals, the committee was well aware of the fact that they would change further and therefore presented "moving targets" for the study. It is for this reason that the committee's recommendations regarding any potential legislation in this area are offered as guiding principles, rather than as specific language for a specific bill.

4. In any new legislation with a fixed term of protection, require database rights holders to provide notice of expiration of the term of protection. Specifically, any such legislation should:

a. Require database rights holders to identify the date on which a database was created so that the user will know when it no longer enjoys statutory protection (of course, those databases that remain commercially valuable longer than the statutory period of protection can continue to be protected by other means, such as copyright, trade secret, contract, and technical and other measures); and

b. Require rights holders of databases that are updated continuously, or at periodic intervals, to identify with reasonable precision those substantial portions of the database that are and are not subject to protection. Failure to identify the date of creation for each new substantial portion of a database should serve as a basis for a defense against infringement after the expiration of the term of protection for the original portion of the database.

5. Apply protection only to databases created after the effective date of any new legislation, in recognition that a major purpose of enacting enhanced protection is to provide additional incentives for the development of new databases.

6. Expressly continue to provide legal rights of access to and uses of proprietary databases equivalent to those that not-for-profit researchers, educators, and other public-interest users enjoyed under traditional or customary practice prior to enactment of any new legislation. Courts should be allowed to invalidate any non-bargained[9] licensing terms that are shown to interfere unduly with otherwise legislatively permitted customary uses by not-for-profit entities.

7. Provide either for a sunset provision with the possibility to renew, or for periodic assessments of the effects of new statutory database protection on competition in the database market and on consumers of databases, as well as on access to and use of data—including S&T data—by not-for-profit, public-interest users, in order to enable timely and appropriate revision of legislation as needed.

8. Although private-sector databases derived from government data should be eligible for protection, protection should not be extended to databases collected or maintained by the government. Any new legislation should

---

[9] By "non-bargained term" the committee means any term, usually contained in a standard form contract, over which, as a practical matter, no actual bargaining by the parties to the contract takes place.

expressly affirm the need for continuation of existing legal norms for wide distribution of government data and of data created pursuant to a government mandate or funding.

Recommended Policy Actions for Government

Although the committee believes that its recommended actions in these areas ought to be undertaken whether or not any new statutory database protection is enacted by Congress, all of these actions will take on an increased urgency and importance if relatively strong proprietary rights are established by federal statute.

1. **Scientific and technical data owned or controlled by the government should be made available for use by not-for-profit and commercial entities alike on a nonexclusive basis and should be disseminated to all users at no more than the marginal cost of reproduction and distribution, whenever possible. While the private sector's creation of derivative databases from government data should be encouraged, the source of the original government data must ensure that those data remain openly available. Any information product derived from a government database also should be required to carry an identifier stating the government source(s) used.**

2. **Federal funding agencies should require university and other not-for-profit researchers or their employing institutions that use federal funds, wholly or in substantial part, in creating databases not to grant exclusive rights to such databases when submitting them for publication or for incorporation into other databases.**

3. **The Copyright Office should sponsor discussions between the representatives of private-sector producers of databases and user stakeholder representatives from government agencies and not-for-profit groups to help develop a common understanding and optimal terms for the licensing of S&T databases and data products.**

4. **Federal government agencies, including federal science agencies as appropriate, should undertake and fund external research that investigates the changing and complex economic aspects of S&T database activities,** particularly in the context of any new legislative database protection measures that may be enacted and in support of legislative principle number 7 recommended above.

5. **All departments and agencies of the federal government should continue to adopt international S&T agreements that include provisions to facilitate access to S&T data across national boundaries and should conduct**

periodic reviews of international policies and agreements to promote confor-mity to the above principles.

6. The U.S. government should negotiate with the Commission of the European Communities to revise its highly protectionist E.U. Database Directive.

Recommended Approach for the Not-for-Profit S&T Community

The not-for-profit S&T community should continue to promote and adhere to the policy of full and open exchange of data at both the national and international levels.

1

Importance and Use of Scientific and Technical Databases

Modern technology has propelled us into the information age, making it possible to generate and record vast quantities of new data.[1] Advances in computing and communications technologies and the development of digital networks have revolutionized the manner in which data are stored, communicated, and manipulated. Databases, and uses to which they can be put, have become increasingly valuable commodities.

The now-common practice of downloading material from online databases has made it easy for researchers and other users to acquire data, which frequently have been produced with considerable investments of time, money, and other resources. Government agencies and most government contractors or grantees in the United States (though not in many other countries) usually make their data, produced at taxpayer expense, available at no cost or for the cost of reproduction and dissemination. For-profit and not-for-profit database producers (other than most government contractors and grantees) typically charge for access to and use of their data through subscriptions, licensing agreements, and individual sales.

Currently many for-profit and not-for-profit database producers are concerned about the possibility that significant portions of their databases will be copied or used in substantial part by others to create "new" derivative databases. If an identical or substantially similar database is then either redisseminated broadly or sold and used in direct competition with the original rights holder's database, the rights holder's revenues will be undermined, or in extreme cases,

---

[1] Box 1.1 provides definitions of *data* and of several other key terms used in this report.

14

Box 1.1
Definitions of Key Terms Used in This Report

Data are facts, numbers, letters, and symbols that describe an object, idea, condition, situation, or other factors. A data element is the smallest unit of information to which reference is made. This report is concerned primarily with digital data, although a large portion of raw data is recorded as analog data, which also can be digitized. For purposes of this report the terms *data* and *facts* are treated interchangeably, as is the case in legal contexts.

Data in a database may be characterized as predominantly *word oriented* (e.g., as in a text, bibliography, directory, dictionary), *numeric* (e.g., properties, statistics, experimental values), *image* (e.g., fixed or moving video, such as a film of microbes under magnification or time-lapse photography of a flower opening), or *sound* (e.g., a sound recording of a tornado or a fire). Word-oriented, numeric, image, and sound databases are processed by different types of software (text or word processing, data processing, image processing, and sound processing).

Data can also be referred to as *raw, processed,* or *verified.* Raw data consist of original observations, such as those collected by satellite and beamed back to Earth, or initial experimental results, such as laboratory test data. After they are collected, raw data can be processed or refined in many different ways. Processing usually makes data more usable, ordered, or simplified, thus increasing their intelligibility. Verified data are data whose quality and accuracy have been assured. For experimental results, verification signifies that the data have been shown to be reproducible in a test or experiment that repeats the original. For observational data, verification means that the data have been compared with other data whose quality is known or that the instrument with which they were obtained has been properly calibrated and tested.

Digital data may be processed or stored on various types of media, including magnetic (RAM, hard drive, diskettes, tapes) and optical (CD-ROM, DVD) media. Data can be made accessible either through portable media or, increasingly, online.

A *database* is a collection of related data and information—generally numeric, word oriented, sound, and/or image—organized to permit search and retrieval or processing and reorganizing. A *data set* is a collection of similar and related data records or data points. Many databases are a resource from which specific data points, facts, or textual information are extracted for use in building a derivative database or data product. A *derivative database,* also called a *value-added* or *transformative* database, is built from one or more preexisting database(s) and frequently includes extractions from multiple databases, as well as original data.

A *database producer* acquires data in raw, reduced, or otherwise processed form—either directly, through experimentation or observation, or indirectly, from one or more organizations or preexisting databases—for inclusion in a database that the database producer is generating. Such database creators—sometimes known as database publishers or originators but for the purpose of this report referred to as database producers—traditionally are the *rights holders* of the intellectual property rights in the databases.

continued

Box 1.1 Continued

In general, *database production* covers all aspects of preparation, processing, and maintenance; development of software for search, retrieval, and manipulation; and documentation of the software and database features and functions prior to distribution of the database by a vendor. Among the wide variety of functions encompassed by database production, in addition to data acquisition, are data reduction (where needed), formatting, enhancing, expanding, merging with other data or data records, categorizing, classifying, indexing, abstracting, tagging, flagging, coding, sorting/rearranging, putting into tabular form, creating visual representations, updating, and putting into searchable and retrievable form for and use and manipulation by users.

A *database vendor* (variously known as a distributor, online host (mostly in Europe and the United Kingdom), disseminator, or provider) sells, leases, or licenses digitized versions of a database on optical disks (e.g., CD-ROM, DVD), floppy disks, tapes, or downloadable complete databases. Many databases, particularly textual ones, are also based on or provided as hard-copy paper publications. A database producer organization may also serve as a database vendor if it both produces a database and provides online access directly to users or sells, leases, or licenses the database.

For the sake of simplicity, the term *database dissemination* or *distribution* as used in this report includes the concept of making databases available online.

The modifier *scientific and technical* designates the subject matter of the database content in the general areas covered in this report.

---

the rights holder will be put out of business. Besides being unfair to the rights holder, this actual or potential loss of revenue may create a disincentive to produce and then maintain databases, thus reducing the number of databases available to others. However, preventing database uses by others, or making access and subsequent use more expensive or difficult, may discourage socially useful applications of databases. The question is how to protect rights in databases while ensuring that factual data remain accessible for public-interest and other uses.

This report explores issues in the conundrum posed by the need to properly balance the rights of original database producers or rights holders and the rights of all the downstream users and competitors—with the principal focus on the balance of rights between the database rights holders and public-interest users such as researchers, educators, and librarians. In particular, the Committee for a Study on Promoting Access to Scientific and Technical Data for the Public Interest focuses on scientific and technical (S&T) data (with examples drawn primarily from the physical and biological sciences) as an essential consideration in reasoned attempts to balance competing interests in databases.

To broaden the perspective of and enhance cooperation among the various competing interests, and to help ensure an efficient and effective outcome for all, the committee examines the following basic elements in the larger issue at hand:

- Salient characteristics and the importance of S&T databases produced and used in research;
- Impacts of computer technology on the production, distribution, and use of S&T databases;
- Motivations of the various sectors involved in S&T research and the dissemination and use of research results;
- Economic issues and incentives that influence the production, distribution, and use of S&T databases, and how these activities are interrelated;
- Mechanisms currently in place for protecting these economic incentives; and
- New legislation currently under consideration that would affect the production, dissemination, and use of S&T databases in a variety of ways.

To ensure the most successful outcome in the current debate over rights in databases, any new action must take account of and balance the legitimate interests of the various stakeholders, and must reflect awareness of how the broad public interest can best be served.

SCIENTIFIC AND TECHNICAL DATA AND THE CREATION OF NEW KNOWLEDGE

Factual data are both an essential resource for and a valuable output from scientific research. It is through the formation, communication, and use of facts and ideas that scientists conduct research. Throughout the history of science, new findings and ideas have been recorded and used as the basis for further scientific advances and for educating students.

Now, as a result of the near-complete digitization of data collection, manipulation, and dissemination over the past 30 years, almost every aspect of the natural world, human activity, and indeed every life form can be observed and captured in an electronic database.[2] There is barely a sector of the economy that is not significantly engaged in the creation and exploitation of digital databases, and there are many—such as insurance, banking, or direct marketing—that are completely database dependent.

Certainly scientific and engineering research is no exception in its growing reliance on the creation and exploitation of electronic databases. The genetic sequence of each living organism is a natural database, transforming biological

---

[2] See Paul F. Uhlir (1995), "From Spacecraft to Statecraft: The Role of Earth Observation Satellites in the Development and Verification of International Environmental Protection Agreements," *GIS Law*, Vol. 2, p. 1.

research and applications over the past decade into a data-dependent enterprise and giving rise to the rapidly growing field of bioinformatics. Myriad data collection platforms, recording and storing information about our physical universe at an ever-increasing rate, are now integral to the study and understanding of the natural environment, from small ecological subsystems to planet-scale geophysical processes and beyond. Similarly, the engineering disciplines continually create databases about our constructed environment and new technical processes, which are endlessly updated and refined to fuel our technological progress and innovation system.

Basic scientific research drives most of the world's progress in the natural and social sciences. Basic, or fundamental, research may be defined as research that leads to new understanding of how nature works and how its many facets are interconnected.[3] Society uses the fruits of such research to expand the world's base of knowledge and applies that knowledge in myriad ways to create wealth and to enhance the public welfare.

New scientific understanding and its applications are yielding benefits such as the following:

- Improved diagnosis, pharmaceuticals, and treatments in medicine;
- Better and higher-yield food production in agriculture;
- New and improved materials for fabrication of manufactured objects, building materials, packaging, and special applications such as microelectronics;
- Faster, cheaper, and safer transportation and communication;
- Better means for energy production;
- Improved ability to forecast environmental conditions and to manage natural resources; and
- More powerful ways to explore all aspects of our universe, ranging from the finest subnuclear scale to the boundaries of the universe, and encompassing living organisms in all their variety.[4]

SCIENTIFIC AND TECHNICAL DATABASES AS A RESOURCE— THE CURRENT CONTEXT

The committee's January 1999 Workshop on Promoting Access to Scientific and Technical Data for the Public Interest: An Assessment of Policy Options,[5]

---

[3] See John A. Armstrong (1993), "Is Basic Research a Luxury Our Society Can No Longer Afford?" Karl Taylor Compton Lecture, Massachusetts Institute of Technology, October 13.

[4] National Research Council (1997), *Bits of Power: Issues in Global Access to Scientific Data*, National Academy Press, Washington, D.C., p. 18.

[5] See online, National Research Council (1999), *Proceedings of the Workshop on Promoting Access to Scientific and Technical Data for the Public Interest: An Assessment of Policy Options*, National Academy Press, Washington, D.C., <http://www.nap.edu>.

included presentations on and discussions of data activities in twelve selected organizations representing three broad sectors (government, not-for-profit, and commercial). The sample activities illustrated some of the depth and range of uses for S&T databases today (Table 1.1 provides a summary) and indicated also the complexity of the often overlapping relationships and interests of database users and producers.

The discussion below outlines basic aspects of current data activities, including collection and production of S&T data and databases, dissemination, and use, and it describes the roles that the three sectors play in the overall process. In contrasting past and current practices, it indicates how ongoing technological advances have contributed to increased capabilities for obtaining and using S&T data. This description, which provides essential background for the remainder of this report, draws on examples from the four general discipline areas—geographic and environmental, genomic, chemical and chemical engineering, and meteorological research and applications—focused on in the workshop.

Collection of Original Data and Production of New Databases

Sources of Primary Data and Uses

The process of scientific inquiry typically has begun with the formulation of a working hypothesis, based usually on limited observation and data, followed by experimentation designed to test the hypothesis. The experimentation results in the accumulation of new data used to confirm or refute the original hypothesis. Understanding of the natural and physical world has been advanced by researchers building on a growing base of knowledge that is continually being refined, tested, and augmented in the long-established approach to scientific inquiry known as the scientific method.

With the advent of digital technologies has come a dramatic increase in the pace and volume of data acquisition. Ongoing rapid advances in electronic technologies for computing and communications, experimentation, and observation ranging from high-frequency direct sampling to multispectral remote sensing have enabled dramatic increases in the quantities of data generated about the natural world at scales from the microcosm to the macrocosm. For instance, the volume of data on weather and climate stored in the National Climatic Data Center has increased 750-fold in the past two decades (Box 1.2). A pharmaceutical company that 5 years ago could characterize 100,000 compounds per year can now handle a million compounds in a week.

Although some of these data represent actual measurements, large quantities of data also are being generated through numerical simulations performed on supercomputers. Collection of new data is becoming increasingly automated as recording devices and instrumentation become more sophisticated and rapid. Moreover, many older paper-based data sets, such as historical U.S. Weather

TABLE 1.1   Examples of Different Types of S&T Database Activities
Discussed in the January 1999 Workshop

Organization (Sector)	Information and Tools Provided	Data Sources
Geographic and Environmental		
U.S. Geological Survey (USGS) (Government)	Geographic data: maps and map products Data from other programs: biologic, geologic, hydrologic	USGS, other federal agencies, state and local governments, not-for-profit researchers, partnerships with private sector
Long-Term Ecological Research (LTER) Network Office (Not-for-Profit)	Site description database, integrated climate database, remotely sensed ecological data	Ecological researchers at distributed sites belonging to the LTER network
GeoSystems Global Corp. (Commercial)	Digital maps, MapQuest Web site, mapping services	From the public domain: government-produced maps (federal, state, local), digital geographic data, remotely sensed imagery Other sources: commercial and other countries' maps, digital data, remotely sensed imagery, other published sources
Genomic		
National Center for Biotechnology Information (Government)	GenBank: DNA and protein sequence data; Other genomic mapping databases; 3D protein structure database; bibliographic databases; software tools	Direct contributions from scientists; access to other databases from government, not-for-profit, other country sources
Center for Bioinformatics University of Pennsylvania (Not-for-Profit)	Specialized biological databases; software tools for integration of distributed heterogeneous databases	Proprietary and public-domain experimental data from academic researchers; manual processing and encoding of data from published literature; online molecular and cellular biology and genomic databases
Molecular Applications Group (Commercial)	Software for storing, mining, and visualizing genomic data; databases derived from public and private data and proprietary software	>150 online database sites, public and proprietary

Users	Dissemination Modes
USGS, other government agencies, commercial database providers and value adders, researchers, and the general public	Maps: hard copy (paper, plastic, film) and digital form; distributed by agency directly and through partnerships with private sector, not-for-profit sector
Researchers	Internet, some tape and CD-ROM for portability
Commercial clients: large companies and consumers	Maps: hard copy and digital form Software products distributed via retail channels (CD-ROM) and directly to corporate customers Mapping services distributed via Internet
Research scientists in academic, government, commercial organizations	Internet access via Web servers and File Transfer Protocol (FTP)
Research scientists in academic, government, commercial organizations (U.S. and abroad)	Internet access Source code distributed directly
Research scientists in academic, government, commercial organizations (U.S. and abroad)	Some software products downloaded from the Web; others require on-site expert installation

continued

TABLE 1.1   Continued

Organization (Sector)	Information and Tools Provided	Data Sources
Chemical and Chemical Engineering		
National Institute of Standards and Technology (NIST) Physical and Chemical Properties Division (Government)	Specialized chemistry and chemical engineering databases (extensively evaluated and documented)	Experimental results from published literature; experiments done specifically for data acquisition; published data evaluations; supplementary data deposits
Chemical Abstracts Service American Chemical Society (Not-for-Profit)	Chemical Abstracts: bibliographic database Registry: registry of chemical substances Software access tools	Journals, patents, books, proceedings, dissertations
Institute for Scientific Information (Commercial)	Bibliographic databases: citation indexes, tables of contents Information services Linkages to publishers' full-text databases	Journals, books, proceedings (print and electronic format)
Meteorological		
National Climatic Data Center (Government)	Climatological summaries from National Weather Service stations; historical long-term climatic databases	National Weather Service, World Meteorological Organization, NASA, bilateral agreements with other countries
Unidata Program, University Corporation for Atmospheric Research (Not-for-Profit)	Quasi-real-time atmospheric and related data Case study data sets Software tools	Public: National Weather Service, National Environmental Data Service Private: network of lightning sensors, sensors in commercial aircraft
TASC (Commercial)	Real-time weather information	Public: National Weather Service—downlink directly from U.S. and international weather satellites, other observational sources

NOTE: Although the subject matter of this study included all S&T databases, the committee was able to choose only representative examples for discussion and analysis in the report. For instance, specific examples from the social sciences or the space sciences, among other disciplines, were not included.

Users	Dissemination Modes
Researchers in academic, government, commercial organizations (some databases used primarily by industrial users)	Variety of forms: hard-copy publication, CD-ROM or floppy disk, Internet access; NIST distributes directly or via agreements with secondary distributors
Researchers in academic, government, commercial organizations; patent examiners; students	Electronic access, hard copy, CD-ROM
Academic, government lab, and corporate libraries; researchers in academic, government, commercial organizations	Diskette, CD-ROM, FTP files, Internet access, hard copy
Individuals, commercial clients, government agencies, engineering uses	Hard-copy, microfiche, magnetic tape, disks, CD-ROM, FTP, Internet
Academic departments	Internet
News media (broadcast and cable TV), aviation, energy and power, agribusiness	Public and private data communication networks: satellite broadcasting services and Internet

**Box 1.2
Example of Large-Scale Data Collection Activity
by the Federal Government**

Statistics from just one discipline in the natural sciences—atmospheric physics—illustrate the explosive growth in the size of some digital scientific and technical databases. The National Climatic Data Center (NCDC) is responsible for storing national, as well as some global, weather and climatic information. Once, most
of these data came from human observations of the current state of the weather,
using simple and straightforward instrumentation, including such commonplace
devices as thermometers, barometers, wind vanes, and rain gauges. The comparatively recent deployment of satellites, sophisticated Doppler radars, lightning detection networks, automatic surface-observing platforms, and heavily instrument-
ed buoys in the marine environment, all linked together through broadband,
high-speed communication systems, has increased the types and volumes of data
collected. The NCDC's storage requirements for these data have increased con-
comitantly by many orders of magnitude. In the period between 1980, when some
of the high-resolution data were just beginning to be recorded, and 1994, when
much of the Doppler radar and lightning data had yet to be generated, the volume
of data stored at the NCDC increased from approximately 1 terabyte to 230 terabytes. By 1999, the NCDC's data holdings had grown to 750 terabytes and are
projected to expand to more than 20 petabytes by 2014. These data are archived
indefinitely and made available to the public.

SOURCE: Information provided by Gerald Barton, National Oceanic and Atmospheric Administration, Washington, D.C.

Bureau observational records or U.S. census data, are being digitized and organized into electronically accessible databases. This shift from a data-poor to a
data-rich research and education environment is occurring through the activities
of a host of government agencies, universities, and other research establishments,
both public and private, nationally and internationally, in diverse research disciplines.

In many cases data are being collected not to answer specific scientific
questions, but rather to describe various physical and biological phenomena in
ever-increasing detail. This broad-based acquisition of data, coupled with data
mining and knowledge discovery[6] and the broad review and analysis of information stored in large databases, is anticipated to reveal trends or patterns or to lead

---

[6] *Data mining* and *knowledge discovery* are related, frequently confused terms, as are *data, information*, and *knowledge*. In the context of electronic databases, the data stored therein remain as data
until they are extracted (mined) and recompiled (put in a context), at which point they become
information. After "information is developed into a collection of related inferences, the data, now

to discoveries that will serve as a source of new hypotheses. The increasing use of databases as a research tool, whether in pursuit of new information or clues to unexpected relationships as starting points for conducting fundamental research, or for developing new commercial applications,[7] relies on the production and availability of such databases as an initial step in the process.

In recent decades, and in most disciplines today, the federal government and federally funded research have played the major role in generating primary S&T data. Substantial amounts and varieties of data are created by thousands of federal government grantees doing basic research, either individually or in teams, and most often at universities and other not-for-profit research institutions. The National Science Foundation and the National Institutes of Health, which in FY 1999 funded over $2.6 billion[8] and $11.8 billion[9] in extramural grants, respectively, provide the bulk of support for these efforts. However, other federal departments and independent agencies also have significant research grant programs that involve the collection of research data and the production of associated databases outside the direct control of the government.

In FY 1998, the federal government spent approximately $19.5 billion on intramural and extramural basic research and almost $50 billion on applied research.[10] A substantial fraction of that funding was devoted to the creation of primary data used for fundamental research, education, and other public-interest purposes. Among the current major observational data research programs are NASA's Earth Observing System and numerous space science missions.[11] The Human Genome Project of the National Institutes of Health is another large-scale, data-intensive research effort.[12] Large experimental facilities dedicated to

---

information, become knowledge." The automated process of evaluating data and finding relationships is data mining, and that of extracting information, especially predicted relationships, or discovering previously unknown patterns among data is knowledge discovery. Numeric databases are more amenable to data mining than textual databases. See Walter J. Trybula (1997), "Data Mining and Knowledge Discovery," pp. 197-229, *Annual Review of Information Science and Technology*, Vol. 32, Martha E. Williams, ed., published for the American Society for Information Sciences by Information Today, Inc., Medford, N.J.

[7] For a discussion of different types of scientific data and their distinguishing characteristics, see National Research Council (1997), *Bits of Power*, note 4, pp. 49-57.

[8] Office of Management and Budget (1999), "Budget of the United States Government, Fiscal Year 2000—Appendix," U.S. Government Printing Office, Washington, D.C., p. 1062.

[9] Office of Management and Budget (1999), note 8, p. 441.

[10] Intersociety Working Group (1999), "Table I-11. Total U.S. R&D 1996-1998," in *Research and Development FY 2000*, American Association for the Advancement of Science, Washington, D.C., p. 71.

[11] See the NASA Web site at <www.nasa.gov> for a description of these major projects under the Earth Sciences Mission home page at <http://www.earth.nasa.gov/missions/index.html> and the Space Science Missions home page at <http://spacescience.nasa.gov/missions/index.html>, respectively.

[12] See the NIH Web site at <www.nih.gov> and the National Human Genome Research Institute home page at <http://www.nhgri.nih.gov/HGP>.

enabling advances in fundamental physics are operated by the Department of Energy[13] or by universities or not-for-profit organizations under contract to one or more federal government agencies.

The government also collects large amounts of data for operational, non-research applications, such as daily weather forecasting, public health and safety, and other public-interest government functions. Many of the resulting databases—such as those developed from observations made by meteorological satellites and ground-based NEXRAD radars operated by the National Oceanic and Atmospheric Administration, or the geological, hydrological, and ecological data collected by the U.S. Geological Survey in response to Department of the Interior mandates[14]—have multiple uses, as well as value for both immediate and long-term research.

Additional extensive data are collected continuously at the state and local government levels, principally in support of public government functions, such as the provision of local health, education, and welfare services, or the regulation of various economic activities. These databases also provide a wealth of factual and statistical information for social science researchers, as well as for historians.

While original data collection activities in the United States, especially for research and educational purposes, are carried out largely under government auspices, a significant amount of basic research is also funded outside government by both not-for-profit and commercial institutions. In 1998, nongovernmental sources spent approximately $15 billion on basic research,[15] some of which was used to produce and analyze new S&T databases. In addition, most large federal government research projects and programs involve one or more foreign government agencies, often with significant international participation of researchers.[16] Large-scale research in areas such as climate trends, marine biology, and space science requires international cooperation in the collection, production, and dissemination of observational data. The effectiveness of such cooperation is dependent on, among other things, agreement on laws and policies for sharing and using those data in different countries.

Significant Aspects of Database Production

The rate of scientific progress depends not only on the collection of new data, but also on the quality of the data collected, their ease of use, and the dissemination of information about the database. Considerable attention must be

---

[13] See the DOE Web site at <www.doe.gov> and the Office of High Energy and Nuclear Physics home page at <http://www.er.doe.gov/production/henp/henp.html>.

[14] See Table 1.1 and the committee's online *Proceedings*, note 5, at Chapter 3, "Characteristics of Scientific and Technical Databases."

[15] Intersociety Working Group (1999), note 10, p. 71.

[16] See National Research Council (1997), *Bits of Power*, note 4, pp. 58-61.

given to all those activities necessary to organize raw or disparate data into databases for broader use. These functions and methods typically include digitally processing the data into successively more highly refined and usable products; organizing the data into a database with appropriate structure, format, presentation, and documentation; creating the necessary accompanying analytical support software; providing adequate quality assurance and quality control; announcing the availability of the database; and arranging for secure near-term storage and eventual deposit in an archive that preserves the database and enables continued access.[17] As databases become ever larger and more complex, effective database production methods become increasingly important and constitute a significant component of the overall cost of the database.

The production of S&T databases requires at least some involvement by those responsible for collecting the original data. Typically, those closest to the collection of the data have the greatest expertise and interest in organizing them into a database whose contents are both available to and readily usable by others. Furthermore, the highly technical and frequently esoteric nature of S&T databases is likely to require that the original data collectors (or project scientists) participate in at least the initial stages of organizing, documenting, and reviewing the quality of data in the database. Involvement by the original data collectors in managing that part of the database production process decreases the probability that unusable or inaccurate databases will result, reduces the need for subsequent attempts to rescue or complete such data sets, and saves time and expense overall.

The level of processing and related database production activities is a significant factor in defining the ultimate utility (and legal protection) of a digital data collection. It is the original unprocessed, or minimally processed, data that are usually the most difficult to understand or use by anyone other than the originator of those data, or an expert in that particular area. With every successive level of processing, organization, and documentation, the data tend to become more comprehensible and easier to use by the nonexpert. As a database is prepared for more widespread use with the addition of more creative elements, it also tends to become more copyrightable as well as more generally marketable. In the case of observational sciences, it is the raw, noncopyrightable data that are typically of greatest long-term value to basic research (see "The Uniqueness of Many S&T Databases," below). Increased or new protection for noncopyrightable databases previously in the public domain could therefore have a disproportionate impact on the heretofore unrestricted access to and use of raw data sets for basic research and education.

Although the production of many S&T databases is performed by, or with

---

[17] See generally, National Research Council (1995), *Preserving Scientific Data on Our Physical Universe: A New Strategy for Archiving Our Nation's Scientific Resources*, National Academy Press, Washington, D.C.

the active participation of, the originating researchers, it also is common for third parties to be involved in an aspect of database production referred to as "value adding." Because the comprehensive production of very large or complex databases can be quite expensive, organizations that collect data, especially in government, are increasingly "outsourcing" database production and subsequent distribution to third parties in an effort to contain costs. In such instances, the raw, or minimally processed, data are provided to a private-sector vendor expressly contracted with by government to add value to the data and produce a database in a commercially marketable format to meet broad user requirements. However, since most federal government databases are openly available and in the public domain, adding value to them may be undertaken as an initiative by entrepreneurs that see a business opportunity in such activities, without any formal contractual arrangement with the government data source. (For examples of such third-party providers, see the summary in Table 1.1 and the committee's online workshop *Proceedings*.)

In the context of this report, the most significant aspect of these third-party, value-adding arrangements is that they almost always involve the transfer of public or publicly funded data and databases to private-sector proprietary database producers and vendors. To the extent that these transfers are done on an exclusive basis and the original government databases are not maintained or otherwise made publicly available, the result is a concomitant decrease in the public availability of S&T data.

Perspective on Number of Databases Produced—Some Statistics

According to one set of recently compiled statistics,[18] over the period from 1975 through 1998 the number of all databases grew by a factor of 38 (from 301

---

[18] The statistics provided in this section were all compiled by Martha E. Williams (1998), "State of Databases Today: 1999," in *Gale Directory of Databases*, L. Kumar, ed., Gale Research, Farmington Hills, MI. Dr. Williams notes the following caveats:

There are undoubtedly large numbers of government and private databases that are not included in these numbers. The government seems not to have a systematic way of making the existence of numeric databases known so that they could be identified and described in database directories where the public could learn about them. The compiler of these statistics estimates that there are tens of thousands of such numeric databases and that the numbers of records contained therein is in the petabyte range. Many such databases are in the hands of individual researchers, some of whom would be reluctant to fill out questionnaires or who consider the data to be of interest only to a small number of known colleagues. The statistics reported herein relate to publicly available databases where the producer wants to make the data publicly known versus databases that are available to the public in theory only. Databases that would require a Freedom of Information Act request or need-to-know are not included in these statistics.

to 11,339), the number of database producers increased by a factor of 18 (from 200 to 3,686), and the number of vendors grew by a factor of 23 (from 105 to 2,459). In 1975 the 301 identified databases contained about 52 million records, whereas in 1998 the 11,339 tallied databases held nearly 12.05 billion records, a 231-fold increase in the number of records.

Although in today's digitized information world databases are produced on all continents, the percentage of all types of databases produced in the United States continues to represent the lion's share of the global output. In 1998, of the 11,339 databases that were identified, 63% were produced in the United States. In 1975, of the 301 publicly available, computer-readable databases worldwide, 59% were U.S. databases. From 1985 to 1993, the ratio of U.S. to non-U.S. databases remained at about 2:1. From 1994 on, production of non-U.S. databases has accelerated somewhat, so that in 1998 the ratio of the number of U.S. to non-U.S. databases was about 3:2. The average size of U.S. databases in terms of the number of records they contained was larger than that of the non-U.S. databases. As noted above, however, most U.S. government and academic databases are not represented in these figures.

In the source quoted here, database statistics were compiled in eight major subject categories—business, health/life/medical sciences, humanities, law, multidisciplinary, news/general, science/technology/engineering, and social sciences. If the health/life/medical sciences category is combined with science/technology/engineering, that general scientific and technical category had the largest number of databases (28%) in 1998, followed by business (26%), news/general (15%), and law (11%), with the remaining three categories accounting for the other 20%.[19]

The Uniqueness of Many S&T Databases

A key characteristic of original S&T databases is that many of them are the only one of their particular kind, available only from a single source, which has significant economic and legal implications, as discussed in subsequent chapters of this report. For example, many S&T databases describe physical phenomena or transitory events that have been rendered unique by the passage of time. Measurements of a snowstorm obtained with a single radar observation, or a statistical compilation of some key socioeconomic characteristics such as income levels collected by a state agency, cannot be recaptured after the original event. The vast majority of observational data sets of the natural world, as well as all unique historical records, can never again be recreated independently and are thus available only as originally obtained, frequently from a single source. Other S&T databases are de facto unique because the cost of obtaining the data was

---

[19] Williams (1998), "State of Databases Today," note 18, p. xxvi.

extremely high. This is the case with very large facilities for physical experiments or space-based observatories.

Even when data similar but not identical to original research results or observations are available for use in non-technical applications, scientists and engineers will likely not find an inexact replica of a database a suitable substitute if it does not meet certain specifications for a particular experiment or analysis. For example, two infrared sensors with similar spatial and spectral characteristics on different satellites collecting observations of Earth may provide relatively interchangeable data products for the non-expert consumer, but for a researcher, the absence of one spectral band can make all the difference in whether a certain type of research can be performed. Thus a database generally deemed adequate as a substitute in the mass consumer market very likely will not be usable for many research or education purposes.

Dissemination of Scientific and Technical Data and the Issue of Access

S&T data traditionally were disseminated in paper form in journal articles, textbooks, reference books, and abstracting and indexing publications. As data have become available in electronic form, they have been distributed via magnetic tape and, more recently, optical media such as CD-ROM or DVD. The growing use of the Internet has revolutionized dissemination by allowing most databases to be made available globally in electronic form. Digitization and the potential for instant, low-cost global communication have opened tremendous new opportunities for the dissemination and utilization of S&T databases and other forms of information, but also have led to a blurring of the traditional roles and relationships of database producers, vendors, and users of those databases in the government, not-for-profit, and commercial sectors. In fact, virtually anyone who obtains access to a digital database can instantly become a worldwide disseminator, whether legally or illegally.[20]

Two of the most important mechanisms for the dissemination of public and publicly funded databases have been government data centers and public libraries. Government, or government-funded, data centers have been created in recent decades for dissemination of data obtained in certain programs or research disciplines. Examples of such data centers include the National Center for Biotech-

---

[20] Of course, this same development is occurring with other forms of online information and proprietary publications outside the S&T database context, such as with copyrighted digital music and videos. For an extensive discussion of the impact of the Internet on various types of information and related intellectual property rights management, see Computer Science and Telecommunications Board, National Research Council (2000), *The Digital Dilemma: Intellectual Property in the Information Age*, National Academy Press, Washington, D.C., in press.

nology Information and the National Climatic Data Center (Table 1.1), but many others have been established for almost every field of research.[21]

Public libraries, whether part of the federal depository library program, university research libraries, or other public libraries or foundations specializing in various S&T or other academic subjects, not only preserve and publicly disseminate government data, but provide general public access for many proprietary S&T databases as well. With ever-increasing costs, however, the libraries' ability to provide this public "safety net" for all published products is diminishing.[22]

Historically, most federal government S&T data and government-funded research data in the United States have been fully and openly available to the public.[23] This has meant that such data are available free or at low cost for academic and commercial research—and indeed any other use—without restrictions and can be incorporated into derivative databases, which can, themselves, be redistributed and incorporated into additional databases. In some instances in which the government contracts for the dissemination of data, however, the rights assigned to the database vendor may place restrictions on the ability of the research and education communities to fully utilize the data. Increasingly, both government and not-for-profit organizations are exploring means to recover database production and distribution costs, or to generate revenue streams in order to support their expensive data activities, thereby making them function in a manner similar to commercial organizations.

The ability to access existing data and to extract and recombine selected portions of them for research or for incorporation into new databases for further distribution and use has become a key part of the scientific process by which new insights are gained and knowledge is advanced. When the ability to access or distribute data on an international basis is required, various intergovernmental agreements are depended on to facilitate such exchanges in the public sector. In contrast, to achieve a suitable return on their investment, private-sector vendors of proprietary databases typically seek to control unauthorized access to and use of their databases. It is at the intersection of public and private interests in data

---

[21] See National Research Council (1995), *Preserving Scientific Data*, note 17, and the accompanying *Study on the Long-term Retention of Selected Scientific and Technical Records of the Federal Government: Working Papers,* National Academy Press, Washington, D.C., and National Research Council (1997) *Bits of Power*, note 4, for a description of many of these government S&T data centers.

[22] As the prices of many serial journal subscriptions substantially outpace the rate of inflation, for example, research libraries increasingly need to rely on interlibrary loans to obtain access for their students and professors. See Association of Research Libraries (1999), *ARL Statistics: 1997-98*, Martha Kyrillidou et al., eds., Association of Research Libraries, Washington, D.C.

[23] As defined in National Research Council (1997), *Bits of Power*, note 4, p. 15, "full and open" availability of data means that "data and information derived from publicly funded research are made available with as few restrictions as possible, on a nondiscriminatory basis, for no more than the cost of reproduction and distribution."

where the greatest challenges emerge. As an example, Box 1.3 sketches some of the issues and approaches currently being tried.

Use of Scientific and Technical Databases

Prior to its public dissemination, the use of a database is limited to those involved in the collection of data or production, and therefore does not provide the opportunity to contribute broadly to the advancement of scientific knowledge, technical progress, economic growth, or other applications beyond those of the immediate group. It is only upon the distribution of a database that its far-reaching research, educational, and other socioeconomic values are realized. One or more researchers applying varying hypotheses, manipulating the data in different ways, or combining elements from disparate databases may produce a diversity of data and information products. The contribution of any of these products to scientific and technical knowledge might well assume a value far greater than the costs of database production and dissemination. The results of a thorough

---

Box 1.3
Database Production in Competitive Research and the Question of Access

Genomic sequence databases exemplify the tension over rights in data and their uses associated with the development of original databases that have both important fundamental research uses and great potential for applied commercial products. Advances in molecular biology and automated DNA sequencing technology have made possible the rapid sequencing of genomes from a variety of life forms, including human beings. These databases are being produced simultaneously by researchers at government, not-for-profit, and commercial laboratories.

Although the government and not-for-profit genomic database producers may be slower than the commercial sector in compiling gene sequence data on the same organisms, they are striving to create analogous databases in order to provide the results on an open basis as a public good for broad research and other uses. Government and not-for-profit sequence data are collected and integrated into major sequence databases in a cooperative international effort that includes the National Center for Biotechnology Information in the United States,[1] the European Molecular Biology Laboratory in the United Kingdom on behalf of the European Union,[2] and the DNA Database of Japan.[3] These centers not only collect and share the data on a daily basis, but also provide some quality control, documentation, and organization of the data before making the information freely available to the scientific and technical community, typically over the Internet. The Human

continued

Box 1.3 Continued

Genome Project aims to provide full sequence data for the human genome and to serve as the future reference standard.

Because of the high intrinsic commercial value of human genomic information for the identification of disease markers and therapeutic agents, commercial entities simultaneously seek to be first in generating primary genomic data, which they can license to pharmaceutical or biotechnology companies, or patent, if possible, to gain market advantage.

While the human genome provides the basic blueprint for human life, it is small differences in individual genes that are likely to provide insight into important questions such as variations in disease susceptibility in different populations (for example, why certain groups of people are predisposed to high blood pressure, diabetes, or Alzheimer's disease). These can be studied by comparing the gene sequences of different populations, such as those individuals susceptible to a disease compared to those individuals who are not. Hence, over time, gene sequence databases of a wide variety of discrete populations will be developed supported by a mix of public and private funding.

Recently, for instance, the Icelandic government formed a controversial partnership with a private U.S. firm to develop a database that will contain genetic information on the entire Icelandic population. Icelanders belong to a highly homogeneous gene pool, which will simplify the detection of disease-related genes. The government gave the firm, by statute, an exclusive license to create and operate that database.[4]

Another recently begun effort involves a consortium of ten U.S. and foreign pharmaceutical companies, together with government and not-for-profit organizations, formed to generate a map of human single-nucleotide polymorphisms,[5] which can be thought of as a low-resolution indicator highlighting areas of variability in the genetic code associated with genetic differences between individuals. Although the research is funded in large part by the commercial sector, the results will be made publicly available. In addition to the cost-sharing benefits of this consortium, a major reason for its establishment is the fear that an individual company, or group of companies, could generate scientifically valuable databases and information on a proprietary basis, preventing broad access and capturing a high proportion of the associated intellectual property rights.[6]

---

[1] See the National Center for Biotechnology Information Web site at <www.ncbi.gov>.

[2] See the European Molecular Biology Laboratory Web site at <www.embl.uk>.

[3] See the DNA Database of Japan Web site at <www.ddbj.nig.ad.ip>.

[4] See J. Gulcher and K. Stefansson (1999), "An Icelandic Saga on a Centralized Healthcare Database and Democratic Decision Making," *National Biotechnology*, Vol. 17, July, p. 620, and Martin Enserink (1998), "Physicians Wary of Scheme to Pool Icelanders' Genetic Data," *Science*, Vol. 281, August 14, pp. 890-891.

[5] See Eliot Marshall (1999), "Drug Firms to Create Public Database of Genetic Mutations," *Science*, Vol. 284, April 16, pp. 406-407.

[6] See Marshall (1999), note 5.

database analysis may reveal a value of the data not apparent in even a detailed examination of the individual elements of the database itself. With the wide-spread availability of information on the Internet have come abundant opportunities to search for scientific and technical gold in this ore of factual elements. The possibilities for discovery of new insights about the natural world—with both commercial and public-interest value—are extraordinary.

In considering how databases are used, it is important to distinguish between end use and derivative use. End use—accessing a database to verify some fact or perform some job-related or personal task, such as obtaining an example for a work memo—is most typical of public consumer uses. End use does not involve the physical integration of one or more portions of the database into another database in order to create a new information product. A derivative (value-adding or transformative) use (see Box 1.1) builds on a preexisting database and includes at least one, and frequently many more, extractions from one or more databases to create a new database, which can be used for the same, a similar, or an entirely different purpose than the original component database(s).

Integration of Distributed Data to Broaden Access and Potential for Discovery

In seeking new knowledge, researchers may gather data from widely disparate sources. A significant advantage arising from the abundance of digitized data now accessible through both private and public networks is the potential for linking data in multiple (even thousands of) databases. The ability to link sites on the World Wide Web is one type of integration that could result in more data being available overall to users. Another is the merging of databases of the same or complementary content. It is now possible to maintain a site with continuously verified links to related information sites for use by subscribers or members of a specific group; an example is the Engineering Village of Engineering Information, Inc.[24] Yet another type of integration occurs in the connection of distributed databases such that different parts of a single large database may reside on different computers in geographically dispersed locations throughout the country or the world. With a common structure, data can be located in a physically distributed network and accessed as if they were in one database in one computer in one location. The cost can thus be distributed and the value of each contributory database increased. Still other databases are automatically created from other databases. For example, data are routinely mined and collected by "knowbots" and "web crawlers" (software employing artificial intelligence and rule-based selection techniques) on the Internet throughout the world and retrieved for pro-

---

[24] See, for example, the Engineering Village of Engineering Information Web site at <www.ei.org/aivillage/village.serve-page?p=4011>.

cessing and further use. One such data-mining activity in the area of biotechnology was described and discussed at the committee's January 1999 workshop (see the Molecular Applications Group's activities summarized in Table 1.1).

With a capability to integrate information in multiple databases comes the potential for exploiting relationships identified in the information and developing new knowledge. In many scientific fields, the initial investment by the database rights holder may not produce the greatest value until it is integrated with the investments of others. For example, while protein sequence data are valuable in their own right, their value is greatly enhanced if associated x-ray crystallographic data are also concurrently available. It is possible to use the combined data to understand the way in which protein chains are folded and, in the case of an enzyme, the way in which various nonsequential residues, or even residues on separate protein chains, combine to form an active site.

Derivative Databases and New Data-driven
Research and Capabilities

The ethos in research is that science builds on science. The creation of derivative databases not only enables incremental advances in the knowledge base, but also can contribute to major new findings, particularly when existing data are combined with new or entirely different data. The importance for research and related educational activities of producing new derivative databases cannot be overemphasized.[25] The vast increase in the creation of digital databases in recent decades, together with the ability to make them broadly and instantaneously available, has resulted in entire new fields of data-driven research.

For example, the study of biological systems has been transformed radically in the past 20 years from an experimental research endeavor conducted in laboratories to one that relies heavily on computing and on access to and further refine-

---

[25] As noted by Vinton Cerf, senior vice president at MCI WorldCom, Inc. ("ACM Awards Keynote," Association for Computing Machinery, New York City, May 15, 1999):

> Scientific databases are proving to be non-linear accelerators of research in specific fields such as biology, astronomy, meteorology, space physics, chemistry, economics, epidemiology, environmental studies and a wealth of other fields. The non-linearity comes about because as each research adds more material to the database, the information is placed in juxtaposition with all other items in the system, exhibiting the same kind of non-linear impact that placing computers in a common network has had, in accordance with Metcalfe's Law (which says that the value of the network grows as the square number of devices in the net). Cerf's Law says that shared databases grow in value in accordance with the number of combinations of data items in the database. When the hundreds of thousands of databases on the Internet and other networks are accessible remotely and can be reached in parallel, and when the partial results can be combined and searched anew, the value of these data can grow dramatically.

ment of globally linked databases.[26] Indeed, one of the fastest growing disciplines is bioinformatics, a computer-based approach to biological research. New technologies, such as DNA microarrays and high-throughput sequencing machines, are producing a deluge of data. A challenge to biology in the coming decades will be to convert these data into knowledge.[27]

The availability of global remote-sensing satellite observations, coupled with other airborne and in situ observational capabilities, has given rise to a new field of environmental research, Earth system science, which integrates the study of the physical and biological processes of our planet at various scales. The large meteorological databases obtained from government satellites, ground-based radar, and other data-collection systems pose a challenge similar to that mentioned above for biology, but also already have yielded a remarkable range of commercial and non-commercial value. Dissemination of the atmospheric observations in real time or near-real time for "nowcasts" and daily weather forecasts has very high commercial value, which is captured by third-party distributors. Use of these atmospheric observations to develop numerical models that predict the weather accurately, hours or days in advance, adds value in terms of safety and economic benefits to society that are not readily quantifiable. While the economic value of these data can be gauged by the profits of private-sector distributors, how does one measure the value of the lives and property saved by timely and accurate hurricane forecasts and tornado warnings? Once the immediate and most lucrative commercial value is exploited, the resulting data continue to have significant commercial and public-interest uses indefinitely. For instance, these data enable basic research on severe weather and long-term climate trends and provide various retrospective applications for industry. The original databases are archived and made available by the National Climatic Data Center (see Box 1.2). Derivative databases and data products are distributed under various arrangements by both commercial and not-for-profit entities like the Unidata Program of the University Corporation for Atmospheric Research (see Table 1.1 and the online workshop *Proceedings*).

Geographic information systems that integrate myriad sources of data provide an opportunity for new insights about the natural and constructed environment, greatly enhancing our knowledge of where we live and how we affect our physical environment. Important applications include environmental management, urban planning, route planning and navigation, emergency preparedness

---

[26] See generally, Working Group on Biomedical Computing (1999), "The Biomedical Information Science Initiative," National Institutes of Health, June 3, available online at <www.nih.gov/welcome/director/060399.htm>.

[27] See Sylvia Spengler and Manfred D. Zorn (1999), "Handling Data Sets in Biology," Lawrence Berkeley National Laboratory Colloquium, Washington, D.C.

and response, land-use regulation, and enhancement of agricultural productivity, among many others.[28]

Finally, databases used by researchers and educators also frequently are produced and disseminated primarily for other purposes. For example, a physical scientist studying the complex relationships among geology, hydrology, and biology as they relate to the preservation of species diversity likely would draw on numerous digital and hard-copy databases originally gathered for other purposes. A social scientist studying the characteristics and patterns of urban crime or the spread of communicable diseases likely would do the same. For many scientists, the ability to supplement existing databases with further data collection in a seamless web of old and new data is basic to meeting the needs of their specific investigations.

Text Databases and Online Publication

Another type of S&T database not yet discussed, but that is used extensively by the research community, consists primarily of text with data summarized or added as examples. These databases may consist of primary literature (as in the case of full-text databases of journal articles) or secondary literature (as in the case of bibliographic reference databases). Traditionally, this text has been available in print form, with publishers providing peer review, professional editing, indexing and formatting, and other services, including marketing and distribution. Increasingly this information is being provided as text databases with the publishers also providing the systems that allow access to these databases. These value-adding or information repackaging functions are performed by both not-for-profit and for-profit organizations. For example, the not-for-profit American Association for the Advancement of Science, a scientific society, produces a database containing the full text of articles from *Science* magazine, including enhancements to the content that do not appear in the print version.[29] Similarly, the for-profit publisher Elsevier Science produces ScienceDirect, a database containing the full text of its journal articles. Bibliographic reference databases are also produced by government, not-for-profit, and for-profit organizations, such as the National Library of Medicine, Chemical Abstracts Service, and the Institute for Scientific Information, respectively (see Table 1.1 and the online workshop *Proceedings*[30]). Where full-text databases include associated data collec-

---

[28] National Academy of Public Administration (1998), *Geographic Information for the 21st Century: Building a Strategy for the Nation*, National Academy of Public Administration, Washington, D.C.

[29] See *Science* online at <http://www.sciencemag.org/>.

[30] See the committee's online *Proceedings*, note 5, at Chapter 3, "Characteristics of Scientific and Technical Databases."

tions, physical and legal possession of the data collections may be retained by the originator or may pass to the publisher.

As S&T data and results are increasingly digitized and made available online, publishers are seeking access to and inclusion of the underlying data collections on which published articles are based. The intent is not only to provide greater validity and support for published research articles, but also to make their online publications more interesting and useful to the S&T customer base. The ability to link to the underlying databases instantaneously and at different levels of detail adds an entirely new and exciting dimension to scientific publishing and to the potential for new research, but also raises the question of who will have the rights to exploiting those data.

THE CHALLENGE OF EFFECTIVELY BALANCING PRIVATE RIGHTS AND THE PUBLIC INTEREST IN SCIENTIFIC AND TECHNICAL DATABASES

The general advancement of knowledge independent of its eventual societal benefits is a goal of basic research. Nevertheless, an endless array of examples demonstrates how the creation of new knowledge, building on the existing base of understanding and information developed by researchers, has enabled broad and important socioeconomic benefits for the nation as a whole. Our society appreciates that knowledge itself is intrinsically valuable and important, and our success in the world market for advanced technology products and services attests to the direct economic benefits of the resulting applications. It is for these reasons that government funds basic research and related data activities as a public good.[31,32] Yet it is precisely these activities that are at risk of being hindered, if not in some instances stopped, by proposed major changes to the legal protections of factual databases.

---

[31] As Lester Thurow points out: "A successful knowledge-based economy requires large public investments in education, infrastructure, and research and development." He goes on to say that

> private returns are apt to be more certain if one is looking for an extension of existing knowledge rather than for a major breakthrough; thus private firms tend to concentrate their money on the development end of the R&D process. Time lags are shorter, and in the business world speed is everything. Because of this proclivity in the private sector, government should focus its spending on the long-tailed projects for advancing basic knowledge. This is where the private firms won't invest, but it is precisely where the breakthroughs that generate business opportunities are made.

(Lester C. Thurow (1999), "Building Wealth: The New Rules for Individuals, Companies, and Nations," *Atlantic Monthly*, June, p. 64.)

[32] For a discussion of public goods in the context of basic scientific research and related data activities, see National Research Council (1997), *Bits of Power*, note 4, pp. 111-114. This issue is discussed in greater detail in Chapters 3 and 4.

Legislative efforts are currently under way in the United States, the European Union, and the World Intellectual Property Organization to greatly enhance the legal protection of proprietary databases. These new legal approaches threaten to compromise traditional and customary access to and use of S&T data for public-interest endeavors, including not-for-profit research, education, and general library uses. At the same time, there are legitimate concerns by the rights holders in databases regarding unauthorized and uncompensated uses of their data products, including at times the wholesale commercial misappropriation of proprietary databases.

Because of the complex web of interdependent relationships among public-sector and private-sector database producers, disseminators, and users, any action to increase the rights of persons in one category likely will compromise the rights of the persons in the other categories, with far-reaching and potentially negative consequences. Of course, it is in the common interest of both database rights holders and users—and of society in general—to achieve a workable balance among the respective interests so that all legitimate rights remain reasonably protected.

2

Incentives and Disincentives Affecting the Availability and Use of Scientific and Technical Databases

As noted in Chapter 1, scientific and technical (S&T) databases are not just a by-product of research, but also an essential foundation on which progress in science builds. Increasingly, too, databases are a research tool that can be sold or licensed and used as the input for new products, the creation of customized derivative databases, and innovations to broaden the scope and increase the pace of discovery. The incentives and disincentives to provide databases reflect all these uses, as well as financial motives. This chapter points out the divergent objectives of organizations that produce and distribute S&T databases, and it outlines some of the economic factors affecting access to such data. In addition, it elaborates on the committee's conclusion that any new legislation that would change the status quo must take into account how it would alter the incentives to produce both original and derivative databases, how it would affect the dissemination and use of databases (especially regarding whether access would be exclusive or unrestricted, particularly for the scientific community), and what the unintended consequences might be for scientific inquiry and other public-interest uses.

DIVERGENT OBJECTIVES OF ORGANIZATIONS THAT PRODUCE AND DISTRIBUTE SCIENTIFIC AND TECHNICAL DATABASES

Original and derivative S&T databases are produced by government, not-for-profit/academic, and for-profit organizations (see Table 1.1 in Chapter 1 for selected examples). Whereas individual researchers typically are motivated by curiosity, a desire to contribute to the knowledge base, and an opportunity to influence the thinking of others, their employers also have in mind a combination of organizational mission and funding considerations. Table 2.1 summarizes typical objectives of government, not-for-profit/academic, and for-profit organizations in producing and disseminating S&T databases and the different weight each places on mission versus financial considerations.

In government and not-for-profit research organizations, including universities, basic research institutes, and national laboratories, the advancement of knowledge as an intrinsic good and in the service of national goals motivates the production and distribution of S&T databases; exploiting data for financial gain is subordinate to fulfilling public-interest objectives. The data products of not-for-profit and government organizations are judged primarily by criteria that are not directly profit related, such as their value to end users, their potential value in advancing a field, their ability to enhance the status of an institution and its research or educational capabilities, and similar objectives typically associated with public-interest or public-good activities related, for example, to improving knowledge of disease factors or interdependencies within ecosystems.

Of course, not all not-for-profit institutions behave in this generalized way. At one end of the spectrum are organizations that, especially if they are fully subsidized, distribute their data freely on the Internet without any effort at cost recovery. Many individual researchers or academics certainly operate this way. At the other end are not-for-profit institutions that seek to maximize the revenues from their S&T databases, subject to the constraints of their tax-exempt status, to finance future R&D and database development in order to remain at the forefront of their respective fields. Most not-for-profits, however, fall somewhere in the middle in trying to reconcile their public-interest mission, on the one hand, with the need to generate sufficient operating revenues, on the other. Universities present a good example of this dichotomy, with the trend in recent years toward greater cost recovery[1] and greater attention to the protection and exploitation of their intellectual property.[2]

In contrast, the for-profit sector seeks mainly to generate profit for management and shareholders. Of course, market success also depends on creating value for users—otherwise, the data products would not be successful. High value can translate to high prices, and such pricing inevitably restricts access. Nevertheless, there are exceptions to the rule here as well, since not all for-profit entities attempt to charge as much as they could for their proprietary databases, perhaps

---

[1] For a discussion of the trend in academic institutions to protect their research results as intellectual property, see Kenneth W. Dam (1998), "Intellectual Property and the Academic Enterprise," John M. Olin Law & Economics Working Paper No. 68 (2d Series), University of Chicago Law School.

[2] Intellectual Property Task Force (1999), "Intellectual Property and New Media Technologies: A Framework for Policy Development and AAU Institutions," Association of American Universities, May 13, Washington, D.C., 31 pp., available online at <www.tulane.edu/~aau/AAUPolicy.html>.

TABLE 2.1 Typical Objectives of Organizations That Produce and
Disseminate S&T Databases

	Federal Government
Primary motivations	Serve national goals, including promoting societal well-being and supporting basic research and other public-good interests
Goals of S&T data collection and database development	Support agency mission; undertake basic research as a basis for economic growth and productivity and for public well-being
Goals of S&T database distribution	Maximize the downstream benefits of basic research; promote availability and use of research results in both public and private sectors
Access to data	Open exchange of information encouraged by federal policy
Interest in protecting the databases produced	Very low; any restrictions generally seen as a problem, with few benefits

NOTE: This table provides broad generalizations regarding the organizations of the three sectors. The committee recognizes that many exceptions and nuances exist, as discussed in this chapter.

instead using such databases as a marketing tool for other products or services, or choosing revenue-generating methods such as advertising as an alternative to charging users for access to data. Such exceptions, however, usually are seen in non-scientific database markets that have large user clienteles.[3]

---

[3] See generally Computer Science and Telecommunications Board, National Research Council (2000), *The Digital Dilemma: Intellectual Property in the Information Age*, National Academy Press, Washington, D.C., in press.

Not-for-Profit/Academic	For-Profit
Fulfill mission, including furthering research, education, creation of knowledge, and discovery; remain economically viable	Achieve corporate objectives, including profit making and growth, and ensure shareholder and customer satisfaction
Advance knowledge by conducting new research and by validating and building on the research of others; educate future researchers; contribute to basis for producing social benefits; build reputation and status of researchers and their institutions	Support development of new or improved products or services; develop databases for direct sale or lease as products or as services in support of other products or services
Encourage open sharing of ideas; enable existing data to be reused for discovery of new knowledge; invite review and validation of research results; facilitate use of research results for product development by S&T community and commercial concerns; recover costs or generate revenue in support of mission	Disseminate data to protect competitive advantage when databases are used for development of other products or are themselves products or services; disseminate via sale or license to generate revenue, enhance customer base and market position, gain competitive advantage, achieve profits, or recover costs
Open, with data and ideas shared after results have been published	Internal and confidential, or available/ marketed externally at a cost set by the organization
Moderate; ranges from very low (for fully subsidized databases) to moderate (when cost recovery is necessary) to high (when data are a source of revenue required to support mission)	Very high; databases regarded as investments to be protected whether they are used in product development or are themselves products or services to be sold

SCIENTIFIC AND TECHNICAL DATABASE COSTS, PRICING, AND ACCESS

Despite their differences in mission and motivation, organizations in all three sectors are constrained by financial responsibilities: federal government agencies must justify their expenditures to Congress; not-for-profit entities must make their organizations at least viable (with any excess of income over expenses reinvested in the organization); and commercial firms must answer to shareholders. All organizations therefore give careful consideration both to the costs associated with database production and distribution and to the potential for generating revenue.

Production and Distribution Costs

The costs of databases can be categorized as production costs and distribution costs. Database production includes both data collection and database preparation. The cost of data collection can be very high and varies with the size, complexity, and difficulty of obtaining the data. It includes the costs of the observational or experimental instrumentation, deployment and maintenance of that instrumentation for the lifetime of the data collection project, and related infrastructure costs such as initial data storage. Data acquisition costs typically represent the bulk of the costs of a database. In major data collection projects such as those involving remote sensing satellites (e.g., National Oceanic and Atmospheric Administration geostationary or polar-orbiting spacecraft), or in large physics experiments requiring specialized facilities (e.g., high-energy particle accelerators), the costs can easily be in the hundreds of millions, or even billions, of dollars. Database preparation costs are those associated with preparing a database for dissemination, including ensuring data quality and accuracy, and enhancing the utility of the data for users. The cost of these efforts can vary from modest to large, depending on the nature of the database and its intended use. For example, the cost of the labor for abstracting or indexing databases can make their production expensive.

Distribution costs include the cost of making, sending, and billing for physical copies; any additional transaction costs such as those for licensing and related administrative activities; and user-specific costs such as those for database maintenance and specialized handling. Distribution costs tend to be much lower than database production costs, particularly if the Internet is the medium of dissemination and little effort is devoted to marketing or user assistance. In some cases, the distribution costs are simply the costs of copying.

Producing databases for online distribution is more a matter of potential market opportunity and incentives, rather than of potential cost savings. Customers have come to expect online distribution, particularly in the S&T data market in which nearly all users are now connected to networks and are technically sophisticated. For such users, online access can improve a database's accessibility, functionality, and utility. From the producer's perspective, it opens new market opportunities or broadens existing ones, but it is more likely to shift costs than to reduce them. The concern over adequate protection of a rights holder's database products is exacerbated in the online milieu by factors such as the possibility of unauthorized access, duplication of content, and mass redistribution. As discussed in Chapter 3, however, making the information available online also can reduce the opportunities for misappropriation of a database by preventing the user from accessing all the underlying data.

Pricing and Access

Because for most S&T databases the production costs are high relative to the subsequent distribution costs, there is a trade-off between efficient access and cost recovery. An economist would define an efficient-access price for a database as one that is equal to the cost of making the database accessible to one additional (the marginal) user. A higher price inefficiently excludes potential users, whether the data activity is in the public or the private sector. The users' welfare could be enhanced without increasing the burden on taxpayers if users were allowed to buy access to the data at the marginal cost of dissemination.[4]

An efficient-access price will almost never generate revenue that matches the database production costs.[5] Instead, the database must be subsidized in some way. Cost recovery has an equity justification, namely that the database is paid for by the users instead of being subsidized out of general revenue. It also subjects the database to a (weak) market test of whether the value to users exceeds the cost. If not, then revenue cannot exceed costs.

Efficient-access pricing is only feasible in conjunction with some other source of revenue or subsidy, such as taxpayer support (in the case of federal agencies) or endowments or industry or government contract support (in the case of not-for-profits). Federal agencies are required by law to provide efficient access, limiting charges to no more than the incremental cost of data dissemination, and not including the average cost of producing the database.[6] Frequently, the federal agency simply charges for the cost of reproduction and distribution, which can be zero if the database is distributed online.[7]

---

[4] Such a price is efficient whether the database is sold or licensed to end users or those using the data in derivative products. Competition among vendors in the derivative market will keep the price low, thus transferring much of the benefit of the underlying data to the consumer. In both cases, efficient access could be preserved with higher revenue if the provider could distinguish users with a high willingness to pay for access and use from those with a low willingness to pay.

[5] For example, in remote sensing systems the development and maintenance costs of the system are very high per user, but the cost of dissemination is trivial in comparison, especially when done online. Information goods such as databases share the essential feature of public goods, namely that use of the good is nonrivalrous. This means that after the first user is served, the marginal cost of allowing access by each additional user is minimal, and the average cost per user is declining (there are economies of scale). Competitive theory does not extend to pure public goods, and the marketplace will not deliver them efficiently, if at all. In order to cover costs, the price must exceed the cost of supplying the marginal user.

[6] See Office of Management and Budget (1993), Circular A-130, "Management of Federal Information Resources," U.S. Government Printing Office, Washington, D.C., regarding federal government information dissemination practices and policies, which were codified in the Paperwork Reduction Act of 1995, P.L. 104-13, which amended 44 United States Code Chapter 35, effective October 1, 1995.

[7] For a discussion of pricing publicly funded S&T data, see National Research Council (1997), *Bits of Power: Issues in Global Access to Scientific Data*, National Academy Press, Washington, D.C., pp. 124-126. On the other hand, providing access can be more costly than it seems, and users

In contrast to government agencies and some not-for-profit entities, profit-motivated firms must, at a minimum, recover their costs. Of course, for-profit entities seek to generate additional income, sometimes using sophisticated pricing strategies, such as segmenting the market with differentiated products and varying the price according to volume, convenience of delivery, customer support services, documentation, scope of subject matter, geographic coverage, and so on.[8] (See Box 2.1 for examples of marketing models for private-sector S&T data products and services.) Unless commercial providers are much more efficient than government providers, the profit motive leads to higher prices for access than a government agency would ideally charge under its mandate to provide wide access.

Whether a commercial database will be made available to public users (such as university laboratories) on better terms than to commercial firms depends on whether there is competition between those two sectors. If the database contains information whose value to commercial buyers is reduced by academic use, then the vendor will not sell it at preferential rates to academics.

Stronger Statutory Protection and the Incentives for Investment

There is a natural link between cost recovery and protection of databases.[9] If databases can be copied without any legal constraints or otherwise freely acquired by users or competitors, then the rights holders will not recover their costs and hence will have no incentive to produce databases.[10] Stronger statutory protection might help prevent unauthorized copying, particularly in unprotected digital formats, and thus promote cost recovery and improve profit margins. In this way it could provide incentives for the creation of new databases in the private sector, especially by commercial entities. However, although this motivation sounds compelling, it should be tempered by the realization, elaborated in

---

must sometimes invoke the Freedom of Information Act to obtain data from the federal government. One interpretation is that the cost of providing the information includes technicians' time, which cannot be disentangled from other activities and is hard to pass on to the user as a cost of dissemination.

[8] See National Research Council (1997), *Bits of Power*, note 7, p. 125.

[9] The trade-off between access and cost recovery is expressed by Richard Gilbert in Chapter 4 of the committee's online workshop *Proceedings* as a trade-off between access by users and protection of rights holders or vendors, where protection refers to market exclusivity that comes from intellectual property rights. See National Research Council (1999), *Proceedings of the Workshop on Promoting Access to Scientific and Technical Data for the Public Interest: An Assessment of Policy Options*, National Academy Press, Washington, D.C., <http://www.nap.edu>.

[10] Some databases, such as consumer-oriented catalogs, are a way to sell products. These databases are not the committee's concern, however, because firms already have every incentive to provide them and they do not require statutory protection. The committee's concern here is with databases for which the pricing for access is the only source of revenue, e.g., most S&T databases.

Chapter 3, that databases already benefit from many forms of protection that permit rights holders to recover their costs. And even when balanced, new legislation could have unintended, negative consequences that need to be avoided, as discussed in Chapter 4.

Moreover, the participants in the committee's January 1999 workshop did not report any instances in which the current lack of statutory protection had dissuaded them from investing in promising new projects. While most of the not-for-profit and commercial-sector participants noted that their organizations' data had been copied on occasion, such copying was written off as an unavoidable loss and part of doing business (similar to shoplifting in retail stores). Commercial-sector participants noted that in repeated instances of suspected infringement, the practice of issuing "cease and desist" letters normally was effective, but when not, they felt appropriately obliged to terminate the infringer's further access (i.e., to their password-protected online databases).[11] In no case, however, had copying of their data prevented the companies represented at the workshop from earning a reasonable profit, over and above full cost recovery.[12]

Mounting Pressures on Government Producers and Distributors of Scientific and Technical Databases for Cost Recovery

Historically, the generation of primary S&T data, in such diverse fields as meteorology, astronomy, and high-energy physics and in the public census, has been funded by governments either directly through specific contracts or indirectly through the sponsorship of academic research. In the United States, the resulting databases of the federal government have been placed in the public domain and have provided the basis for subsequent research.

In many cases in which the government produces databases, it distributes the raw or partially processed data as a public good and lets not-for-profit organizations and commercial firms customize the data for special market segments or individual users. This achieves a better balance between requirements for cost recovery and the advantages to the public of efficient-access pricing. Under this approach, taxpayers subsidize the collection and preservation of the raw or partially processed data, but users pay the entire additional cost of customizing the

---

[11] Several participants reported that pirated editions of their databases were being sold in developing countries. This kind of wholesale copying is illegal under domestic and international copyright and unfair competition laws. The ongoing failure of other nations to respect and enforce existing intellectual property law is largely a concern of new World Trade Organization rules known as the TRIPS agreement (see note 5 in Chapter 3). Increased worldwide protection of databases would require a new treaty, and its effectiveness could depend on its integration into the TRIPS agreement.

[12] An extensive background report prepared in advance of the workshop also failed to uncover any negative consequences for companies. See Stephen M. Maurer, Appendix C, in the committee's online *Proceedings*, note 9.

Box 2.1
Marketing Models for Private-Sector S&T Data
Products and Services

Traditional Marketing Methods

Many of the business plans for creating and distributing databases are similar to those used in the software industry. Referred to here as "traditional" methods, they are usually tailored to the perceived market's size and wealth:

• **Mass-market products.** The typical commercial database directory lists thousands of low-cost, mass-market databases with prices ranging from a few hundred to tens of thousands of dollars each. This approach extends to scientific and engineering products. Examples include POISINDEX (a medical database that links many thousands of poisons to treatment protocols and is aimed at emergency room personnel), Science and Technology Network (STN) International (on-line databases of the physical and mechanical properties of thousands of materials), and the Institute for Scientific Information, Inc. (see Table 1.1 in Chapter 1)

• **Specialty-market products.** Many database producers concentrate on developing products aimed at relatively small, high-value markets. Examples from Table 1.1 include Molecular Applications Group (software for storing, mining, and visualizing genomic data) and TASC (custom weather data for aviation, agribusiness, and power companies). Not surprisingly, such products tend to be costly—indeed, large pharmaceutical and biotechnology firms may pay millions and even tens of millions of dollars per year in licensing fees for access. Sales contracts normally include detailed confidentiality provisions.

• **Custom products.** Large pharmaceutical houses sometimes ask bioinformatic companies to provide exclusive access to a particular database. For example, in 1994 SmithKline Beecham agreed to pay $125 million to Human Genome Sciences, Inc. for exclusive rights to one such proprietary database containing EST (gene fragment) information.[1]

---

[1] Jon Cohen (1997), "The Genomics Gamble," *Science*, Vol. 275, February 7, pp. 767-772.

---

databases, which are prepared and sold mostly by commercial firms. There are problems with both systems of finance, however. As mentioned above, subsidies avoid the market test of whether the willingness to pay for the data exceeds their cost. Commercial provision at higher prices must face this market test, but the higher prices inefficiently restrict access. In the evidence collected by the committee, there was no suggestion that federal agencies were collecting too much raw data, so it is reasonable to price the raw data for efficient access, relying on

New Marketing Methods

In addition to taking a traditional approach to marketing databases, vendors have also devised effective alternatives for selling their products. The following sampling of new methods is provided by way of illustration:

- **Bundling.** Product-linked databases are frequently sold as part of a package that includes other products. For example, medical and scientific instrument makers may bundle their products with relevant nuclear physics data. Databases are also included in the price of some proprietary search tools.
- **Use of market makers.** Many companies create elaborate databases to help users find and use their products. For example, semiconductor chip manufacturers commonly prepare elaborate "cookbooks," "libraries," and design tools to help engineers use their products. Similarly, some online bibliographic services are made available to consumers at little or no cost. When a user's research is successful, such services offer to sell reprints at costs ranging from $10 to $20 per article.
- **Migration of old products to new media.** Some publishers have been able to create new digitized versions of databases as an outgrowth of existing print products. Electronic versions of journals and other print-based resources are probably the classic example. Current vendors include Elsevier Science (ScienceDirect), John Wiley & Sons, Springer-Verlag, and Academic Press. Extensions of the concept, which would link traditional articles to online data sets, are already being developed and implemented.

SOURCE: Commissioned paper by Stephen M. Maurer, "Raw Knowledge: Protecting Technical Databases for Science and Industry," Appendix C in National Research Council (1999), *Proceedings of the Workshop on Promoting Access to Scientific and Technical Data for the Public Interest: An Assessment of Policy Options*, National Academy Press, Washington, D.C., <http://www.nap.edu>.

taxpayer subsidies to the extent possible. Of course, taxing for general revenue also involves some inefficiency.[13]

---

[13] Charging an access price higher than the efficient-access price can be thought of as funding the database through an excise tax. However, basic principles of optimal taxation suggest that income taxes are less distorting than isolated excise taxes. See, for example, Richard Tresch (1981), *Public Finance: A Normative Theory*, Irwin-Dorsey Limited (Georgetown, Ontario), p. 320, who says:

However, because of mounting budgetary pressures generally, and increasing costs of public information management specifically, government database providers worldwide have recently been coming under pressure to recover all costs, rather than only the cost of distribution. This pressure toward full cost recovery results both from the rising costs associated with the rapidly expanding rate of data collection and from foreign pressures. European governments, for example, are turning increasingly to a full-cost-recovery approach for their S&T database production and dissemination activities. The E.U. Database Directive puts considerable pressure on non-E.U. countries, including the United States, to do the same.[14] Although existing law in the United States precludes adoption of such restrictions for government data, the enactment of a strong new database protection statute for private-sector databases could stimulate further interest in privatizing U.S. government database dissemination activities. By making databases more profitable, new protectionist legislation could shift responsibility for their creation from the public sector to the private sector. The social harm of such a shift would be an increase in the price for access, especially for highly specialized databases—such as some S&T databases—with a comparatively small market. A possible social benefit would be that the private sector would be subjected to a weak market test of whether the value of the databases exceeded their cost. As noted above, however, most data collected by public agencies are raw data whose full potential value has not yet been realized, and in many cases, user-oriented transformations of the data are already in the hands of the private and nonprofit organizations that might have a better sense of the user market.

Pressures for cost recovery also arise because the benefits that accrue to consumers and the broader society under the efficient-access pricing model are harder for legislators and administration policy makers to see (and measure) than those that accrue as reduced tax burdens under the cost-recovery model, or those that accrue as increased profits under the commercial model. Thus, science agencies in the United States are increasingly turning to the private sector, to both not-for-profit and commercial entities, in outsourcing government S&T database dissemination activities, or even to purchase data from commercial suppliers. For example, in order to promote private-sector investment and development of space technologies and applications, the Commercial Space Act of 1998 encourages NASA—an agency engaged to a substantial degree in basic research activities—to purchase space and Earth science data products and services from the private sector, and to treat data as a commercial commodity under federal procurement regulations. The potential negative effects of this trend are discussed in some detail in Chapter 4.

"Income taxes are held in high regard by many public sector economists. . . . (They are) seen as being reasonably efficient, based on large empirical literature which indicates that the supply of labor and capital are both extremely price inelastic."

[14] See Stephen M. Maurer (1999), "Raw Knowledge: Protecting Technical Databases for Science and Industry," Appendix C in the online workshop *Proceedings*, note 9.

3

Access to and Protections for Databases:
Existing Policies and Approaches

The escalating drive to enhance legal protection for databases arises primarily from three developments. The first is the evolution of a digital world in which information is an increasingly important commercial commodity whose unauthorized appropriation can be accomplished cheaply and accurately, and the information broadly disseminated. The second is the 1991 U.S. Supreme Court decision in *Feist Publications, Inc. v. Rural Telephone Service Co.*,[1] limiting copyright protection to creative works and denying protection to a "sweat-of-the-brow" database whose composition, even though it may require the investment of effort and resources, is not sufficiently creative in selection and arrangement to qualify for copyright protection. The third is the European Parliament's adoption in 1996 of the Directive on the Legal Protection of Databases (hereinafter, the E.U. Database Directive; the E.U. Directive)[2] that requires countries of the European Union to adopt strong property protection for databases (see Appendix D for the full text of the E.U. Directive). The E.U. Directive also stipulates that *sui generis* protection for databases in Europe will be extended to foreign database rights holders only if their home countries have adopted substantially similar

---

[1] 499 U.S. 340 (1991). It is important to note, however, that *Feist* did not "overturn" the "sweat-of-the-brow" doctrine under copyright, which Congress had actually done already under the Copyright Act of 1976. Moreover, the sweat-of-the-brow doctrine under state law was never a prevailing legal approach.

[2] Directive 96/9/EC of the European Parliament and of the Council of 11 March 1996 on the Legal Protection of Databases, 1996 O.J. (L77) 20. The E.U. Database Directive is reprinted as Appendix D of this report.

protection.[3] These three developments have resulted in a perceived increased vulnerability of databases to misappropriation and to a new European legal regime that has been alleged to place U.S. database rights holders at a competitive disadvantage in Europe.[4] It is the last factor that appears to concern private-sector scientific and technical (S&T) database producers and vendors the most, based on input received at the committee's January 1999 workshop. Nevertheless, it is important to note that all other laws protecting foreign rights holders in the European Union remain independently applicable to them under the national treatment clause of the Agreement on Trade-Related Aspects of Intellectual Property Rights (TRIPS Agreement)[5] and related conventions.

This chapter briefly describes the law and policy governing U.S. government databases; the existing legal, technical, and market-based measures that are available to protect private-sector databases in the United States; and the new E.U. Database Directive.

ACCESS TO U.S. GOVERNMENT-FUNDED SCIENTIFIC AND TECHNICAL DATA

The U.S. government is the world's largest creator, user, and disseminator of data and information, including the federal records and S&T databases that are considered highly valuable national assets. A basic principle underlying most U.S. information law is that democracy thrives and the economic and social benefits of information are maximized in society by fostering wide diversity in the creation, dissemination, and use of information.[6] By extension, to gain the greatest economic and social benefits from government information assets, such information should be made available to all in the most efficient, timely, and equitable ways possible. U.S. laws and policies generally implement this proposition. In direct contrast to those laws that encourage protection of the proprietary rights of private-sector entities, U.S. domestic information policy at the federal level may be summarized as one comprising "a strong freedom of infor-

---

[3] E.U. Database Directive (1996), note 2, Recital 56.

[4] See testimony of Henry Horbaczewski, Reed Elsevier, Inc., on behalf of the Coalition Against Database Piracy in June 15, 1999, hearing on H.R. 1858, the Consumer and Investor Access to Information Act of 1999, before the House Commerce Subcommittee on Telecommunications, Trade, and Consumer Protection, U.S. House of Representatives, U.S. Congress, Washington, D.C.

[5] See Final Act Embodying the Results for the Uruguay Round of Multilateral Trade Negotiations, done at Marrakesh, Morocco, April 15, 1994, reprinted in *The Results of the Uruguay Round of Multilateral Trade Negotiations: The Legal Texts 2-3* (GATT Secretariat ed., 1994); Marrakesh Agreement Establishing the World Trade Organization, Annex 1C: Agreement on Trade-Related Aspects of Intellectual Property Rights, Apr. 15, 1994. The TRIPS Agreement holds all member states of the World Trade Organization to a common set of intellectual property norms.

[6] 44 U.S.C., section 3506(d)(1)(A) (Supp. 1995).

mation law, no government copyright, fees limited to recouping the cost of dissemination, and no restrictions on reuse."[7]

U.S. law expressly forbids federal departments and agencies from claiming copyright in their written works, thereby placing these information resources in the public domain. The 1976 Copyright Act states that "[c]opyright protection under this title is not available for any work of the United States Government."[8] The reasons are several. One is the fundamental belief that government copyright of public records is the antithesis of open access whereby an informed citizenry can check official actions and possible abuses. However, other values also are at work. Taxpayers should not have to pay twice for the same information—once for the cost of generating the work, and a second time to obtain it. Also important to avoid is the danger that government could exercise copyright in a manner that would burden free speech (e.g., so as to prevent critics from obtaining particular information at any price). Finally, individuals ought to be able to derive benefit from public goods (such as public S&T data and information) and enjoy improved educational opportunities through increased access to data and information, opportunities that are inherently beneficial in their own right.[9] Thus, the position of Congress has been to support the development of secondary markets for government information by individuals and private businesses, and to otherwise encourage the distribution of government information in the public interest.

The U.S. Freedom of Information Act (FOIA)[10] and the open records laws of the individual states[11] together balance the right of citizens to be informed about government activities and the need to maintain the confidentiality of some government records. Both the national FOIA and state open records laws generally support a policy of broad disclosure by government. For instance, if a database held by a federal agency is determined to be an agency record, the record must be disclosed to any person requesting it unless the record falls within one of nine exceptions contained in the FOIA.[12] Exceptions are construed nar-

---

[7] Peter N. Weiss and Peter Backlund (1997), "International Information Policy in Conflict: Open and Unrestricted Access versus Government Commercialization," in *Borders in Cyberspace: Information Policy and the Global Information Infrastructure*, Brian Kahin and Charles Nesson, eds., MIT Press, Cambridge, MA.

[8] 17 U.S.C., section 105.

[9] U.S. Congress, Office of Technology Assessment (1986), *Intellectual Property Rights in an Age of Electronics and Information*, U.S. Government Printing Office, Washington, D.C.

[10] 5 U.S.C., section 552.

[11] R. Daugherty, G. Leslie, and L. Reis, eds. (1997), "Tapping Official's Secrets: A State Open Government Compendium," Reporters' Committee for Freedom of the Press, Arlington, VA, available online at <www.rcfp.org/tapping/index.cgi>.

[12] The nine exceptions as set forth in 5 U.S.C., section 552 (b) are as follows:

(b) This section does not apply to matters that are—(1) (A) specifically authorized under criteria established by an Executive order to be kept secret in the interest of national defense or foreign policy and (B) are in fact properly classified pursuant to such Execu-

rowly by the courts so that disclosure is typically favored over non-disclosure. In responding to citizen requests for records, government agencies at most levels are authorized to recover the costs of responding to those requests.

Federal departments and agencies also have affirmative obligations to actively disseminate their information as defined by the provisions of OMB Circular A-130.[13] They are particularly encouraged to disseminate raw content on which value-added products can be based and to do so at cost of distribution and through diverse channels, with no imposition of restrictions on the use of the data. The core provisions of OMB Circular A-130 were incorporated into the Paperwork Reduction Act of 1995,[14] which additionally encourages agencies to use information technology to provide public access, rather than relying on cumbersome FOIA processes. Given federal agencies' expanding use of World Wide Web servers to meet their internal objectives, as well as to better implement the government's data-sharing policies, the additional cost of disseminating information to the public has become so negligible that many government databases are now made freely available to anyone with the ability to access them over the Internet.[15]

---

tive order; (2) related solely to the internal personnel rules and practices of an agency; (3) specifically exempted from disclosure by statute (other than section 552b of this title), provided that such statute (A) requires that the matters be withheld from the public in such a manner as to leave no discretion on the issue, or (B) establishes particular criteria for withholding or refers to particular types of matters to be withheld; (4) trade secrets and commercial or financial information obtained from a person and privileged or confidential; (5) inter-agency or intra-agency memorandums or letters which would not be available by law to a party other than an agency in litigation with the agency; (6) personnel and medical files and similar files the disclosure of which would constitute a clearly unwarranted invasion of personal privacy; (7) records or information compiled for law enforcement purposes . . . ; (8) contained in or related to examination, operating, or condition reports prepared by, on behalf of, or for the use of an agency responsible for the regulation or supervision of financial institutions; or (9) geological and geophysical information and data, including maps, concerning wells. Any reasonably segregable portion of a record shall be provided to any person requesting such record after deletion of the portions which are exempt under this subsection. The amount of information deleted shall be indicated on the released portion of the record, unless including that indication would harm an interest protected by the exemption in this subsection under which the deletion is made. If technically feasible, the amount of the information deleted shall be indicated at the place in the record where such deletion is made.

[13] Office of Management and Budget (1993), Circular A-130, "Management of Federal Information Resources," U.S. Government Printing Office, Washington, D.C.

[14] Paperwork Reduction Act of 1995, P.L. No. 104-13, 109 Stat. 163 (May 22, 1995), 44 U.S.C. Chapter 35.

[15] For example, federal legislation and court decisions generally may be accessed online at <http://thomas.loc.gov/>, while many federal geographic data sets may be accessed online at <http://fgdclearhs.er.usgs.gov/>. The databases being made available by federal agencies typically may be traced through their official Web sites indexed online at <http://lcweb.loc.gov/global/executive/fed.html>.

Open access policies specifically targeted at public and publicly funded S&T data availability and exchange, on agency and interagency program levels as well as internationally, have been adopted in recent years especially in the area of environmental and earth science research.[16] These policies all restate in similar terms the federal policy of full and open availability of data pointed to in Chapter 1.

The Federal Acquisition Regulation (FAR) applies generally to all federal agency contractual relationships with the private sector.[17] In subpart 27.4, "Rights in Data and Copyrights," the FAR delineates the respective rights and obligations of the government and the contractor regarding the use, duplication, and disclosure of data produced under contracts with the government.[18] As a general proposition, the government acquires unlimited rights in most data first produced in the performance of a contract, while the contractor may receive limited rights in some data.

Article 36(a) of OMB Circular A-110, "Uniform Administrative Requirements for Grants and Agreements with Institutions of Higher Education, Hospitals, and Other Not-for-profit Organizations,"[19] states that the recipient of a grant or agreement subject to this circular "may copyright any work that is subject to copyright and was developed, or for which rights holdership was purchased, under an award. The Federal awarding agency(ies) reserve a royalty-free, nonexclusive and irrevocable right to reproduce, publish or otherwise use the work for Federal purposes, and to authorize others to do so." Article 24(h) states, "Unless Federal awarding agency regulations or the terms and conditions of the award provide otherwise, recipients shall have no obligation to the Federal Government with respect to program income earned from license fees and royalties for copyrighted material, patents, patent applications, trademarks, and inventions produced under an award."

Similarly, the *Grant Policy Manual* of the National Science Foundation (NSF) specifies in Section 732.1 that the following principles govern the treatment of copyrightable material produced under NSF grants:[20]

---

[16] For a comprehensive collection of such policies in the environmental data area, see Interagency Data Management Working Group of the U.S. Global Change Research Program (1999), *Data Access Policy Actions of Importance to Global Environmental Change Data Users*, U.S. Global Change Research Program, Washington, D.C. See also, the Data Policies portion of the U.S. Global Change Research Program's Global Change Data and Information System Web site online at <www.gcdis.usgcrp.gov/policies>.

[17] 48 CFR, Chapter 1.

[18] 48 CFR, at subpart 27.4 on "Rights in Data and Copyrights."

[19] Office of Management and Budget (1997), Circular A-110, "Uniform Administrative Requirements for Grants and Agreements with Institutions of Higher Education, Hospitals, and Other Not-for-profit Organizations," revised November 19, 1993; as further amended August 29, 1997.

[20] National Science Foundation (1995), *Grant Policy Manual*, NSF 95-26, Arlington, VA, available online at <www.nsf.gov:80/bfa/cpo/gpm95/start.htm>.

a.   NSF normally will acquire only such rights in copyrightable material as are needed to achieve its purposes or to comply with the requirements of any applicable government-wide policy or international agreement.

b.   To preserve incentives for private dissemination and development, NSF normally will not restrict or take any part of income earned from copyrightable material except as necessary to comply with the requirements of any applicable government-wide policy or international agreement.

c.   In exceptional circumstances, NSF may restrict or eliminate an awardee's control of NSF-supported copyrightable material and of income earned from it, if NSF determines that this would best serve the purposes of a particular program or grant.

Cooperative research and development agreements (CRADAs) provide yet another legal mechanism by which databases or services can be created to meet particular governmental needs.[21]   The CRADA legislation, however, creates an exception to the Freedom of Information Act. Databases created under a CRADA potentially may be withheld from citizen requests under FOIA.[22]

State and local governmental entities in the United States also create and maintain records and databases that have substantial value for various segments of the research and educational community. State governments historically have been a primary source of detailed information in the areas of health, welfare,

---

[21] 15 U.S.C., section 3710a:

As used in this section—(1) the term "cooperative research and development agreement" means any agreement between one or more Federal laboratories and one or more non-Federal parties under which the Government, through its laboratories, provides personnel, services, facilities, equipment, intellectual property, or other resources with or without reimbursement (but not funds to non-Federal parties) and the non-Federal parties provide funds, personnel, services, facilities, equipment, intellectual property, or other resources toward the conduct of specified research or development efforts which are consistent with the missions of the laboratory, ... See 15 U.S.C. 3710a(d)(1).

[22] See *DeLorme Publishing Company, Inc. v. National Oceanic and Atmospheric Administration,* 917 F.Supp. 867 (DC Maine 1996) upholding 15 U.S.C. 3710a as a legislative exception to FOIA:

15 USC 3710a provides two types of protection from disclosure of privileged or confidential information resulting from cooperative research and development activities. First, (n)o trade secrets or commercial or financial information that is privileged or confidential . . . which is obtained in the conduct of research or as a result of activities under this chapter from a non-Federal party participating in a cooperative research and development agreement shall be disclosed. See 15 USC 3710a(c)(7)(A). Second, (t)he director, or in the case of a contractor-operated laboratory, the agency, for a period of up to 5 years after development of information that results from research and development activities conducted under this chapter and that would be a trade secret or commercial or financial information that is privileged or confidential if the information had been obtained from a non-Federal party participating in a cooperative research and development agreement, may provide appropriate protections against the dissemination of such information, including exemption from subchapter II of chapter 5 of Title 5.

education, labor markets, transportation, the environment, and criminal justice.[23] Because communities have a great interest in knowing about themselves and their activities, local governments often produce detailed databases on the characteristics and attributes of physical, social, and human resources in the community that are unavailable from other sources.

The U.S. Copyright Act does not explicitly ban copyright claims in the works of state and local governments, as it does for the works of the U.S. government.[24] As such, most state and local governments believe they have the option of asserting copyright in their public records if they choose to do so. Some legal scholars argue that although allowed by law, generally it is unwise economic and social policy for state and local governments to allow government commercialization of public information.[25] Other legal scholars argue that claims of copyright by state and local governments in many of their works and databases are illegal.[26] Under the patents and copyright clause of the U.S. Constitution, the argument is made that Congress lacks the ability to extend copyright beyond that which is necessary to provide "incentives" to authors to make their works available.[27] When state or local government agencies collect information in response to a legislated obligation, it is the public need as defined by the legislative obligation that provides the incentive to gather information or to create a public record. If copyright failed to exist, the information would still be collected. This being the case, copyright provides no incentive for data collection and database production, and the works therefore may not be protected by copyright.[28] Yet other legal scholars claim that government commercialization of public information raises significant First Amendment free speech issues.[29] One argument is that contractual provisions that ban the reuse or further dissemination of public infor-

---

[23] Weiss and Backlund (1997), note 7, p. 304.

[24] 17 U.S.C., section 105.

[25] In support of the general proposition for all levels of government, see L. Ray Patterson and Craig Joyce (1989), "Monopolizing the Law: The Scope of Copyright Protection for Law Reports and Statutory Compilations," *UCLA Law Review*, Vol. 36, p. 719; J.H. Reichman and Pamela Samuelson (1997), "Intellectual Property Rights in Data?" *Vanderbilt Law Review*, Vol. 50, p. 51; and J. Littman (1992), "After Feist," *U. Dayton Law Review*, Vol. 17, p. 607. For a state-level statement of policy in general accord see Minnesota Government Information Access Council, *Digital Democracy: Citizens' Guide for Government Policy in the Information Age*, available online at <http://www.admin.state.mn.us/ipo/giac/report/index.html>.

[26] H. Perritt (1996), *Law and the Information Superhighway,* John Wiley & Sons, New York, p. 484.

[27] United States Constitution, Article I, Section 8, clause 8.

[28] See generally the discussion in Perritt (1996), note 26, pp. 482-487.

[29] See Henry Perritt, Jr. (1996), *Section 11.10 First Amendment Role, Law and the Information Superhighway,* Wiley Law Publications, New York, pp. 489-491; Philip H. Miller (1991), "Life After Feist: Facts, the First Amendment, and Copyright Status of Automated Databases," *Fordham L. Rev.*, Vol. 60, pp. 507, 509; and Michael J. Haungs (1990), "Copyright of Factual Compilations: Public Policy and the First Amendment," *Colum. J. Law & Soc. Probs.*, Vol. 23, pp. 347, 364.

mation or that establish varying fee structures depending on purpose of use might readily be used by government officials for ulterior motives of censorship or manipulation of public information for political purposes.

Regardless of the legal and economic arguments, many local and state agencies have pursued the imposition of copyright in at least some public records and databases, both hard copy and digital.[30] These local government authorities perceive the possibility of paying for the creation and maintenance of local government information systems other than through general tax revenues. Restricting access to public records is contrary to the plain letter language of most state open records laws in the United States, and therefore explicit legislation is typically required to allow the restrictions. Those who seek to impose new access restrictions on citizens bear the burden of overcoming the underlying policy arguments on which the existing laws are based, foremost of which are that open access keeps government accountable and that open access to government information, subject to appropriate limitations based on privacy, confidentiality, national security, and other considerations, has far greater long-term economic benefits for a community than does pursuing revenue-generation approaches.

It is noteworthy that the United States has become a world leader in research and technology at a time when its domestic public information laws have been so divergent from those of most other nations. In general, the U.S. legal system allows greater access to and use of government information at the local, state, and national government levels than is allowed in other nations. U.S. law also grants individuals greater leeway to use the work products of others without permission than is often granted by the laws of other nations. The role of the U.S. legal system in supporting full and open access to scientific data for the academic and commercial sectors and the role of U.S. federal funding in defraying the costs of collecting and providing access to scientific data are factors that should not be overlooked when exploring the competitive success of U.S. scientists and businesses.

EXISTING PROTECTIONS FOR DATABASES IN THE UNITED STATES

Currently available legal protections of databases in the United States include copyright, private contracts and licensing, trade secret law, and state unfair

---

[30] Iver Petersen (1997), "Public Information, Business Rates: State Agencies Turn Data Base Records Into Cash Cows," *New York Times*, July 14, p. D1; *For What It's Worth: A Guide to Valuing and Pricing Local Government Information* (1996), Public Technology, Inc. Press, Washington, D.C. (Note: Public Technology, Inc. is a not-for-profit technology organization of the National League of Cities, the National Association of Counties, and the International City/County Management Association.) For a survey of local government policies relating to the distribution of digital geographic information, see H.J. Onsrud, J.P. Johnson, and J. Winnecki (1996), "GIS Dissemination Policy: Two Surveys and a Suggested Approach," *Journal of Urban and Regional Information Systems*, Vol. 8, No. 2, pp. 8-23.

competition law. Significant augmentation of the existing legal regime is provided for online databases by various technical means as well as by additional market-based measures.

Legal Protections

Copyright Law

A database can be protected under the Copyright Act as a "compilation," defined as a work that results from the collection and assembly of data that are "selected, coordinated, and arranged" in an original way.[31] As the Supreme Court stated in the *Feist* decision, if the selection or arrangement of the data displays a "modicum of creativity" it is protectable by copyright.[32] The term of copyright protection is long—the life of the author plus 70 years, or in the case of a work for hire, the shorter of 95 years from first publication or 120 years from the year of creation.[33] An unauthorized reproduction, which is not otherwise privileged by the law, is illegal, and substantial civil and criminal remedies exist to punish infringers.[34]

Since the fall of 1998, copyright law has prohibited the manufacture and sale of devices designed primarily to circumvent technologies such as signal scrambling and encryption that are used to protect copyrighted works. Beginning in the fall of 2000, the law will also prohibit the attempt to circumvent such technological protections. The 1998 law, known as the Digital Millennium Copyright Act,[35] contains some exemptions for libraries, educational institutions, law enforcement, and research activities, and opens the possibility of additional future exemptions to be made after further study of the statute's operation by the Librarian of Congress. The new statute also prohibits the removal or alteration of "copyright management information" from copyrighted works. "Copyright management information" essentially means any identifying mark, such as the name and address of an author or copyright rights holder that is associated with a work.[36]

The copyright law, however, permits some unauthorized uses that are deemed to be "fair" or that are specifically exempted from infringement in the statute.[37] Section 107 of the 1976 Copyright Act states that:

---

[31] 17 U.S.C., section 101.

[32] *Feist*, note 1.

[33] 17 U.S.C., section 302.

[34] 17 U.S.C., sections 501-512.

[35] Digital Millennium Copyright Act, P.L. 105-304 (October 28, 1998), U.S. Congress, Washington, D.C.

[36] 17 U.S.C., sections 1201 and 1202.

[37] 17 U.S.C., section 107.

. . . the fair use of a copyrighted work, including such use by reproduction in copies or phonorecords or by any other means specified by that section, for purposes such as criticism, comment, news reporting, teaching (including multiple copies for classroom use), scholarship, or research, is not an infringement of copyright. In determining whether the use made of a work in any particular case is a fair use the factors to be considered shall include—

(1) the purpose and character of the use, including whether such use is of a commercial nature or is for nonprofit educational purposes;

(2) the nature of the copyrighted work;

(3) the amount and substantiality of the portion used in relation to the copyrighted work as a whole; and

(4) the effect of the use upon the potential market for or value of the copyrighted work.[38]

A more significant limitation on copyright protection, particularly for databases, is that copyright protects only the manner of expression and does not "extend to any idea, procedure, process, system, method of operation, concept, principle, or discovery" incorporated into a copyrighted work.[39] For some databases, the line between protected expression and unprotected facts can be difficult to identify. Generally speaking, however, copyright would most likely protect the selection or arrangement of a database, but not the data as such. Consider, for example, a scientific journal article reporting the results of an experiment, with the data from the experiment reproduced as an appendix in a database qualifying as a compilation. Copyright would protect the narrative description of the experiment, but subsequent researchers could use the findings of the experiment without permission. Moreover, other researchers could extract the data from the appendix.

Unlike in Europe and in many other countries, U.S. copyright law does not protect works of authorship created by the federal government.[40] This asymmetry is significant, since a similar asymmetry with the European Union would exist upon enactment of any new database protection legislation in the United States.

Because of the speculative nature of the outcome of basic scientific research, much science is conducted with the use of public funds by researchers in government and in academia who do not directly depend financially on the economic exploitation of their results. The realities of the scientific process and the limited legal protection for data under copyright law have contributed to the tradition and culture within the scientific community of open sharing of ideas and data discussed in Chapters 1 and 2.

---

[38] 17 U.S.C., section 107.

[39] 17 U.S.C., section 102(b).

[40] 17 U.S.C., section 105.

Private Contracts and Licensing

Rights holders of trade secrets may protect the confidentiality of their information by disclosing the information only to those who agree, in a binding contract, to keep it confidential and to refrain from reproducing the information. Where two individuals bargain face to face and one agrees to disclose in return for a promise of confidentiality, a court ordinarily will enforce the terms as it would any other contractual provision. If the information is sold outright to another party—such as when a book is sold to a consumer—contractual restraints on the buyer's right to use or resell the copy usually are not enforced.[41] Increasingly, however, information products are not "sold," but rather are licensed with restrictive terms as to use.[42] For example, all the commercial database vendors who participated in the committee's January 1999 workshop, as well as those not-for-profits that disseminated their data for a fee, indicated that they relied primarily on site licensing arrangements for disseminating their databases to customers. Because customers typically have the opportunity to read the license terms and conditions in advance, and even to make agreed-upon changes, such terms are ordinarily enforceable.

Database rights holders can easily protect their interests with a confidentiality agreement when distribution of a database is limited to those who directly contract with them. The fact that contract terms are only effective between the contracting parties and not binding on third parties who may get access to the database has been cited as a weakness,[43] since many databases must be publicly distributed in order to be commercially viable. In recent years, however, rights holders of some digital databases have marketed their databases to the public using "shrink-wrap" license agreements. These are agreements that are enclosed within the package containing the database (usually a database on a CD-ROM or on other electronic media) and provide notice on the outside of the package that

---

[41] The "first sale" doctrine of copyright law, 17 U.S.C., section 109, specifically authorizes the rights holder of a copy to "sell or otherwise dispose of" the copy. See also *Bobbs-Merrill v. Straus*, 210 U.S. 339 (1908).

[42] For a discussion of the emerging legal issues pertaining to online database and information licensing, see the special issue on licensing of information and proposed changes to Article 2B of the Uniform Commercial Code (UCC) in *Berkeley Technology Law Journal* (1998), Vol. 13, No. 3, and in *California Law Review* (1999), Vol. 87, No. 1. Both present the results of a symposium, "Intellectual Property and Contract Law for the Information Age," held at the University of California, Berkeley, in April 1998. The symposium Web site is at <www.sims.berkeley.edu/BCLT/events/ucc2b/ucc2b.html>. In April 1999, however, the efforts to amend UCC Article 2B were terminated and the proposed revisions to state law in this area were proposed instead as the "Uniform Computer Information and Technology Act." For comprehensive background information on the evolution of this issue, see generally "A Guide to the Proposed Uniform Computer Information Transactions Act" online at <www.2Bguide.com>.

[43] See U.S. Copyright Office (1997), *Report on Legal Protection for Databases*, U.S. Congress, Washington, D.C., available online at <lcweb.loc.gov/copyright/reports>.

breaking the "shrink-wrap," or entering the electronic gateway if online, constitutes acceptance of the terms within. These terms often include restrictions on distribution of the database to others. The enforceability of these provisions remains uncertain, however. Some legal scholars argue that the shrink-wrap license constitutes a valid contract, while others believe that a contract cannot exist unless the parties have access to the terms prior to paying for the product.[44] Similarly, some leading cases have upheld the shrink-wrap license and the terms that restrict use;[45] other cases have refused to enforce the terms of a shrink-wrap license.[46] Most shrink-wrap licenses permit the licensee to return the product, unused, if the terms are unacceptable. The reality is that most consumers do not read these license terms.

Private transactions are an important method of distributing valuable information and are increasingly the method of choice for providing access to data and information on the Internet. On the one hand, particularly with vulnerable digital information, the right to distribute information with restrictions on use allows original rights holders of databases to capture the economic returns from their initial investment. Moreover, private transactions are flexible in permitting the two parties to tailor their agreement to the mix of their particular interests, as long as they have the opportunity to negotiate the terms. On the other hand, enforcing "contractual" terms imposed through shrink-wrap licenses (and now "click-on" licenses in the online medium), which effectively are imposed on the public at large, may interfere with the balance between private property rights and public-interest access to information.[47] A term in a shrink-wrap or click-on license that prohibits what would otherwise be a privileged use of the data might effectively limit scientists' access to or use of the raw materials necessary for their research, contrary to public-interest policies.

Trade Secret Law

State trade secret law protects valuable commercial information that is kept secret by its rights holder from unauthorized access or reproduction by improper means.[48] Trade secret doctrines do not require that the information be kept absolutely secret, but the trade secret rights holder must take reasonable steps to maintain the confidentiality of the information. Trade secret law also might be

---

[44] See generally, Mark Lemley (1995), "Intellectual Property & Shrinkwrap Licenses," *USC L. Rev.*, Vol. 68, p. 1269.

[45] *Pro CD v. Zeidenberg*, 86 F.3d. 1447 (7th Cir. 1996).

[46] *Vault Corp. v. Quaid Software Ltd.*, 655 F. Supp. 750 (1987).

[47] See J.H. Reichman and Jonathan Franklin (1999), "Privately Legislated Intellectual Property Rights: Reconciling Freedom of Contract with Public Good Uses of Information," *U. Penn. Law Review*, Vol. 147, p. 875.

[48] See generally, *Restatement (Third) of Unfair Competition*, sections 39-45 (1995).

applicable in the absence of any contractual provisions for confidentiality if the parties understood that the disclosure was in fact in confidence. Once information is made public it loses its secrecy and enters the public domain. It then can be free for others to use, absent some form of protection, such as copyright.

Unfair Competition Law in State Common Law

When Congress acts in areas in which it has authority (e.g., copyright, interstate commerce), federal law preempts state law when Congress explicitly so states in the legislation, or when state law would interfere with implementation of the federal law. The copyright system preempts state laws that duplicate or disrupt the protection accorded works of authorship by copyright.[49] Thus, states may not grant a general right to database owners to prevent unauthorized reproduction or use of databases that qualify as an original work of authorship, and it is unlikely that they could grant a naked right against reproduction or use to unoriginal databases. However, in some states, a common law cause of action for misappropriation is understood to survive preemption and under limited circumstances may provide protection to some database owners from certain forms of unfair competition.

The doctrine of misappropriation derives from the early U.S. Supreme Court decision in *International News Service v. Associated Press.*[50] In that case, a news wire service appropriated the dispatches of a competing service from published newspapers on the East Coast and published them simultaneously and in direct competition with the originating service on the West Coast. The Supreme Court, while denying that a statutory property right could exist in the news, found that the unauthorized appropriation in this case was prohibited because compiling the data gave rise to a "quasi-property right" and also because the unauthorized appropriation directly undermined the investment in news gathering of the originating service. Courts have usually been reluctant to apply the decision beyond the facts of the case, and the *Restatement (Third) of Unfair Competition,* section 38 (1995) described the misappropriation doctrine as lacking a coherent application.[51]

The most recent discussion of the misappropriation doctrine occurred in *National Basketball Ass'n v. Motorola, Inc.*[52] Although the court recognized a limited scope for the doctrine—one confined to situations in which time-sensitive

---

[49] *Restatement (Third) of Unfair Competition,* section 301 (1995). Cf. *Sears, Roebuck & Co. v. Stiffel Co.,* 376 U.S. 225 (1964); *Compco Corp. v. Day-Brite Lighting, Inc.,* 376 U.S. 234 (1964); and *Bonito Boats Inc. v. Thunder Craft Boats, Inc.* 489 U.S. 141 (1989).

[50] 248 U.S. 215 (1918).

[51] But see *Goldstein et al. v. California,* 412 U.S. 546 (1973) allowing state misappropriation protection of noncopyrightable sound recordings.

[52] 105 F.3d 841 (2d Cir. 1997).

data were appropriated and used in direct competition with the originator—it refused the request of a professional basketball league to prohibit the unauthorized taking and distribution of the scores of its currently played games to paid subscribers of a paging service.

While database rights holders can and do assert protection under the misappropriation doctrine, the reluctance of the courts to apply this doctrine and the likelihood that its broad application would be preempted by the Copyright Act makes the misappropriation doctrine of questionable value in protecting databases beyond those with extremely time-sensitive value, such as real-time stock-price quotations. However, nothing prevents Congress from developing a minimalist form of statutory protection that builds on this foundation.

Technological Protections

The danger of database misappropriation can be mitigated with increasing efficiency by technologies that help enforce the terms of licensing contracts, or that enable the rights holder to keep the database as a trade secret while also providing access to subsets of data at arm's length.[53] A number of technological innovations have been developed to provide various forms of security, privacy protection, and intellectual property management. Table 3.1 provides a summary of some of these approaches. No form of computer protection is perfect, and no method will likely prevent copying of small amounts of data. Moreover, it is almost certain that every technological security method will eventually be able to be countered through the use of other technological advances. The technological

---

[53] The committee did not focus extensively on the increasingly important area of technological protections for digital information, because a concurrent NRC report is examining this issue in depth. See Computer Science and Telecommunications Board, National Research Council (2000), *The Digital Dilemma: Intellectual Property in the Information Age*, National Academy Press, Washington, D.C., in press. For additional information on these technologies, see Mark Stefik and Teresa Lunt, "Overview of Technologies for Protecting and Misappropriating Digital Intellectual Property Rights: The Current Situation and Future Prospects," Chapter 5 in the committee's online *Proceedings*. See National Research Council (1999), *Proceedings of the Workshop on Promoting Access to Scientific and Technical Data for the Public Interest: An Assessment of Policy Options*, National Academy Press, Washington, D.C., <http://www.nap.edu>. Also see *Berkeley Technology Law Journal*, Vol. 13, No. 3, Fall 1998 (issue dedicated to intellectual property and contract law); I. Trotter Hardy (1998), *Project Looking Forward: Sketching the Future of Copyright in a Networked World*, report prepared for the U.S. Copyright Office; Mark Stefik and Alexander Silverman (1997), "The Bit and the Pendulum: Balancing the Interests of Stakeholders in Digital Publishing," *American Programmer*, September, pp. 18-35 (also published in *The Computer Lawyer*, Vol. 16, No. 1, pp. 1-15, January 1999); Brian Kahin and Kate Arms, eds. (1996), *Proceedings: Electronic Commerce for Content*, Forum on Technology-Based Intellectual Property Management, Interactive Multimedia Association, Annapolis, MD; Bruce Schneier (1996), *Applied Cryptography*, second edition, John Wiley & Sons, New York; and Lars Lyberg, ed. (1993), *Journal of Official Statistics*, Special Issue on Confidentiality and Data Access, Vol. 9, No. 2, Statistics Sweden.

TABLE 3.1 Summary of Technological Approaches to Protection of Databases

Approach	What It Does	Domain of Use	Limitations
Encryption	Codes information so that a secret key is needed in order to read it	Any digital data; protects data during storage and transmission	Protection vanishes when data decrypted for use; strong security required for safeguarding keys
Watermarking	Embeds covert tracing or identifying information	Digital photos, music, text, and databases	Can be easy to remove or tamper with; may confound data merging
Download limitations	Limits amount of data that can be downloaded at one time or from one site	Currently deployed; discourages wholesale, automatic downloading of databases from the Web	Presents only a weak barrier to a determined adversary
Database access control	Classifies data in a database into different levels or groups for access, and limits use based on agreed-to set of rules	Intelligence and business organizations in which individuals or different groups have access to different portions of a database according to job function or need to know	More relevant to online databases than to offline scientific databases, or databases of more general interest
Hardware-based trusted system	Enforces a digital contract that governs fees for and uses of digital works	High-security systems used mainly in military and intelligence applications	Not widely deployed; inexpensive hardware augmentation provides only minimal support; outstanding legal issues about liability and enforceability of digital contracts
Software-based trusted system	Supports same functions as hardware-based trusted systems, except that support is provided only by software	In trial use for distribution of video, music, and documents	Vulnerable to being disabled by simple means; subject to wide-scale system failure triggered by computer viruses

approaches that currently are available, however, can hinder or prevent, to varying degrees of efficiency, the wholesale copying and redistribution of databases without compensation to their rightful rights holders. These approaches are reviewed briefly in the discussion below.

Encryption

A powerful and frequently cited technology for computer security is encryption, the encoding of data to make them unreadable to those who do not have the key for deciphering them. Separating those who should have access from those who should not, encryption enables differences in level of security with increases in key length. Encryption is applicable to practically any kind of digital information. It is the technology of choice for protecting data in storage and during transmission over an insecure channel.

But encrypted data are only as secure as the key, and there are different approaches to the problem of protecting the key. Given the data and the key, a knowledgeable user can decrypt the data and then distribute them, or distribute the encrypted data together with the key. If the protections in the system are insufficient, there are various ways in which an attacker could obtain the key or gain access to an unencrypted form of data (e.g., digital music or video while it is being played back, or text or data while they are being displayed). Thus, although encryption is an important element of any modern computer security system, it often must be combined with other elements in a security architecture to achieve the degree of protection desired for the digital data or information product.

Watermarks

The term "watermark" initially signified a special mark made in paper during its manufacture. The mark, which becomes visible when the paper bearing it is held up to a light, is taken as indicating an original. The term now covers a wide range of technologies for embedding information in digital files and rendered works, including text, pictures, and audio. As a technique for intellectual property protection, watermarks carry information that identifies a work or provides a means of tracing its purchaser or user. Watermarks can be visible or hidden. Hidden watermarks are designed to avoid interfering with the use of the data. For example, in digital music, watermarks can be encoded in such a way that they are not detectable to the human ear when the music is played. A watermark could be hidden in a database as extra (but unused) elements of data that would not interfere with information processing systems.

Watermarks do not prevent copying but could potentially provide a means for tracing the source of an unauthorized copy. This trace-back capability provided by watermarks is not necessarily foolproof; those who would misappropriate data or otherwise infringe on rights in a database might be able to write programs for tampering with or removing the watermarks. Such tampering may well be illegal under the recent Digital Millennium Copyright Act, however.

Online Database Access Controls

Several kinds of technology have been developed to control or limit access to databases, particularly those available online. One of the simplest is an online regulator, which limits the quantity of information that can be downloaded by a given individual or site.[54] The technology provides an impediment to automatic data-mining programs that would acquire a database's contents by merging the results of a large set of individual queries. A second approach to online database access control marks the data according to different levels and categories and regulates the availability of the information to authorized parties. This technology is relevant in business and intelligence applications, where access to information is regulated according to "need to know" tiers or categories of access.

Trusted Systems

"Trusted" systems are those that can be relied on to obey certain rules for distributing information. In the context of intellectual property protection, the rules take the form of a digital contract between the information provider and user. The contract spells out fees and other terms and conditions of use, such as the period of time over which the information can be used and whether the user is allowed to print out or make copies of the information for distribution or sale.

Trusted systems with secure hardware support have been used in military and intelligence organizations for several years and currently are in limited use in digital and networked publishing. For example, they support pay-per-view and subscription viewing of satellite television services. Software and hardware for distributing music via trusted systems are in early prototyping and testing stages, and systems supporting digital network document and software publishing are in limited use on personal computers and in "electronic books." Although trusted systems for database access could be developed, such currently available technologies possess minimal security and control measures. The systems seldom have digital contract provisions specifying more than the duration of the subscription and perhaps the number of simultaneous users.

---

[54] For example, several of the commercial and not-for-profit database vendors at the committee's January 1999 workshop noted that this online protection strategy, in conjunction with other available technical measures and well-structured licensing terms, provided satisfactory protection for their businesses, despite the possibility that a determined individual could access and download from a database under multiple identities, or collude with others to do so, in order to reconstruct illegally the entire database.

Several obstacles prevent effective, widespread deployment of trusted systems for databases. First, the legal standing of digital contracts enforced by machines has many practical limits with respect to enforceability and liability. Second, appropriate public-key infrastructure to support authentication and authorization is not yet widely used or available. Third, the only commercially viable approach to trusted systems for databases depends exclusively on software for implementing security, but software-only approaches are vulnerable to tampering and to widespread, catastrophic failure caused, for example, by computer viruses. Fourth, it can be both very costly and time consuming to attain the levels of confidence necessary to achieve a "trusted system," which conflicts with the rapid pace of the industry and time-to-market considerations.

Summary of Technological Protections

Several technological measures have been developed that can be deployed for management of intellectual property rights in databases. The main goal of protection is to prevent widescale unauthorized redistribution of databases without compensation to database rights holders. Although no totally effective technological solution has yet been developed to protect intellectual property comprehensively, several measures are already in use with increasingly satisfactory results. A potentially effective technological approach appears to be the use of trusted systems, with digital contracts that specify appropriate terms and conditions. These systems would use encryption technology for protecting databases during storage or communication, watermark technologies to enable tracing the source of pirated copies when such theft occurs, and database access controls and query governors to flexibly control database access.

Current limitations affecting technology available for protection of ownership rights in databases include absence of a widespread public-key infrastructure for encryption, legal uncertainties about the enforceability of digital contracts, and the relatively low level of security that is possible with software-only security systems. In addition, despite advances in technological measures for protecting digital databases, human fallibility—or overt malicious action—will continue to result in system security breaches for the foreseeable future.

Market-based Database Protection Through Updating and Customizing

There are various business practices that database vendors can use to protect their investments.[55] One type of protection for databases arises from their commercial perishability. Many data become rapidly obsolete; consequently, data-

---

[55] See Computer Science and Telecommunications Board (2000), *The Digital Dilemma*, note 53, in press.

bases are updated frequently. For example, meteorological data and stock market prices are provided in real time on a continuous basis. Some biotechnology databases are updated every night. Most commercially viable databases are updated at least annually. Since copying intrinsically introduces a lag, updating provides some level of protection against piracy, because the copier, like the database originator, must provide updates, thereby reducing the cost advantage of copying.[56]

Frequent updates can constrain the market price for a database because the most recent update competes with its previous version in the market, although the former versions may nonetheless be almost as useful. Database pricing will almost certainly permit recovering the cost of updating, but the cost of the original database might not be recoverable. Collaborations among the government, not-for-profit, and commercial sectors, however, can overcome some of these problems. For example, as discussed at the January 1999 workshop, commercial meteorological, geographic, and biotechnology database producers utilizing the original data made available freely under the federal government's full and open access mandate have successfully marketed and disseminated their value-added databases. The joint effort of the original public-data collector and commercial database value-adder and vendor accomplished the twin goals of enhanced data quality and wide dissemination at a reasonable price. Most important, in the context of this report, this broad distribution of data was achieved without statutory database protection.

Another market-based approach used by database producers and vendors to limit the potential for misappropriation, while meeting the needs of their customers, is production of customized or targeted versions of their databases. Different versions of the same database tailored to different market segments can appeal to a broader swath of the market while making it more difficult for an unfair competitor to steal all versions and undermine the customer base.[57]

Finally, database producers or vendors who have a well-established reputation in the market will have an advantage over most competitors who would copy their products. Customers are frequently willing to pay more to vendors who are reputed to sell quality databases and data products.

TIPPING THE BALANCE:
THE EUROPEAN UNION'S DATABASE DIRECTIVE

Other nations have legal and other protective measures for databases similar to those already in place in the United States, but a discussion of foreign law is

---

[56] See Stephen M. Maurer in Appendix C of the committee's online *Proceedings*, note 53.

[57] See generally, Carl Shapiro and Hal R. Varian (1998), "Versioning: The Smart Way to Sell Information," *Harvard Business Review*, Nov.-Dec., pp. 106-114.

beyond the scope of this study. There is, however, one important new legal development—the aforementioned E.U. Database Directive—that is particularly relevant to the present discussion, since it has been cited by commentators[58] as well as by congressional legislators[59] as a major driver for the adoption of a similar legal regime in the United States (see Appendix D for the full text of the directive).

The E.U. Database Directive requires that each member country of the European Union (and affiliated states) adopt legislation protecting databases.[60] The E.U. Directive imposes a uniform copyright provision that protects only the "selection or arrangement" of the contents of a database that is the "author's own intellectual creation."[61] Countries are permitted to provide for privileged unauthorized uses in accordance with the Berne Convention for the Protection of Literary and Artistic Works.[62] The specific privilege recommended for not-for-profit educational or scientific uses is very narrowly limited, however, "for the sole purposes of illustration for teaching or scientific research as long as the source is indicated and to the extent justified by the non-commercial purpose to be achieved."[63]

The E.U. Directive also provides for an independent right to protect databases that are not protectable by copyright. The right attaches to any database that is a product of substantial investment and prohibits any extraction or reutilization of a substantial part of a protected database—judged qualitatively or quantitatively—without permission of the rights holder.[64] The E.U. Directive provides that a noncopyrightable database is protected for 15 years from its date of completion.[65] "Lawful users" of databases that have been made available to the public may extract or use insubstantial parts of the database for any purpose and may make other such use that does not conflict with the "normal exploitation of the database or unreasonably prejudice the legitimate interests of the maker of the database."[66] Member states may, but are not required to, incorporate some very narrow and specific exceptions, including one for the purposes of illustra-

---

[58] See Reichman and Samuelson (1997), note 25.

[59] See statement by Senator Orrin Hatch on Database Antipiracy Legislation, *Cong. Rec.,* Vol. 106, S. 316 (Jan. 19, 1999).

[60] E.U. Database Directive, note 2, Article 16. The directive required all member states to comply with its requirements by January 1, 1998. Only a few had done so by that date, and not all countries had complied as of September 1999.

[61] Id., Article 3(1).

[62] Id., Article 6.

[63] Id., Article 6(2)(b).

[64] Id., Article 7.

[65] Id., Article 10. Although the nominal term of protection is limited to 15 years, Article 10(3) has the effect of extending protection in perpetuity to databases that continue to be updated or revised pursuant to a "substantial new investment, evaluated qualitatively or quantitatively."

[66] Id., Article 8(2).

tion for teaching or scientific research that is more limited than the one provided for under copyright.[67]

Most significant, from the U.S. perspective, the E.U. Directive provides that member states should make the protection applicable only to databases owned by nationals or habitual residents of a member state or to databases owned by nationals of a third country only if the third country offers comparable protection to databases produced by nationals of a member state.[68]

Although preliminary drafts of the E.U. Database Directive were founded on an unfair competition law model of database protection, the final version was based on a strong property rights model. The initial right to exclude extraction or use applies even when the unauthorized use is not a competitive threat to the protected database. Only express privileged uses can escape potential liability. The privilege for scientific research appears to apply only for non-commercial purposes, and this is further qualified by an ambiguous limitation for the purposes of illustration. Since most scientific research has at least the potential for commercial application, including commercial publication, and is not simply for "illustration," the privilege may turn out to be a very narrow one, indeed, even if it is adopted by a member state. Similarly, the "insubstantial part" exception is undermined by the qualitative impact test. Moreover, the term of the right is 15 years, and potentially much longer, a very long period given the commercial half-life of many kinds of scientific data.[69]

When combined with unrestricted online licensing rights, strong database protection legislation such as the E.U. Directive subjects a research user of, say, a chemical handbook, to a starkly different situation than that experienced under traditional copyright law under the print paradigm. Table 3.2 provides a summary comparison of research user rights under these two legal regimes.[70] The net result of unrestricted licensing coupled with strong statutory database protection is that the most borderline of all the objects of protection under intellectual property law—raw or factual data, whether S&T or any other—paradoxically receives the strongest scope of protection available from any intellectual property regime except, perhaps, patent law.[71] The committee believes that the adoption of a law such as the E.U. Directive, either in the United States or internationally, would retard the advancement of science, the growth of knowledge, and opportunities for innovation.

---

[67] Id., Article 9(b).

[68] Id., Recital 56 and Article 11.

[69] See the discussion of the term of protection in Chapter 4.

[70] See J.H. Reichman and Paul F. Uhlir (1999), "Database Protection at the Crossroads: Recent Developments and Their Impact on Science and Technology," *Berkeley Technology Law Journal*, Vol. 14, No. 2, pp. 799-821.

[71] Reichman and Samuelson (1997), note 25, p. 94.

TABLE 3.2   Summary Comparison of Not-for-Profit Research User Rights
Under Traditional Copyright and Under Online Licensing When Combined
with *Sui Generis* Database Protection Legislation

Traditional Copyright Law	Licensing Plus *Sui Generis* Protection Such as Provided by the E.U. Database Directive
1. User can immediately use all disparate factual data and information disclosed in a database; copyright law does not protect ideas or facts. Fair-use exception available for certain additional research or educational uses, even of protected expression.	1. Even after paying for access to factual data and information, which are not copyrightable by definition, user faces limitations on use in any ways prohibited by the license and as reinforced by the legislation; user cannot distribute another database, using the same factual data or information, without either seeking permission and perhaps paying another fee or regenerating those protected data independently.
2. User can independently create another version of a database and sell it; copyright law allows independent creation, and all factual data are in the public domain.	2. User can independently create another version of the database. If this is not possible, user needs a license or permission to combine legitimately accessed factual data or information into a derivative data product; the licensor can claim that the user is violating redistribution and other rights, and the user must guess what courts will consider to be a quantitatively or qualitatively insubstantial part of the database; the licensor is under no duty to grant such a license; and if a sole source, the licensor may not want any competition from follow-on products.
3. User can combine noncopyrightable factual data with other data and information into a multiple-source or interdisciplinary database for research or educational purposes without permission or additional payment to the originators.[a]	3. User cannot lend, give, or sell data to others even after paying for access (unless permitted by the license) because there is no first sale, only a license; user would have to obtain express permission and perhaps pay additional fees to avoid the risk of harming the market (e.g., possibly causing one lost sale).
4. User can make limited or "fair use" of even protected expression for not-for-profit research or classroom purposes; such uses often deemed fair or privileged uses under statutory law or precedents.	4. Because there are no limits on licensing, user is subject to database vendor overriding even those exceptions contained in the legislation, including exceptions for research, education, or other public-interest uses.
5. Following the first purchase of a copyrightable database in hard copy, user can lend, give, or resell it to anyone else under the first-sale doctrine, borrow it from a library, use it at any time for virtually any [lawful] purpose, and make a copy of it for personal or scholarly purposes.	5. During the period of protection, user rights depend on the terms of the license supported by the new property right; database would not enter the public domain for at least 15 years (and in Europe possibly never if the rights holder continues to invest in maintenance or updates of a dynamic database).

NOTE: This summary table was compiled from a more detailed comparative discussion presented in an article by J.H. Reichman and Paul F. Uhlir (1999), "Database Protection at the Crossroads: Recent Developments and Their Impact on Science and Technology," *Berkeley Technology Law Journal*, Vol. 14, No. 2, pp. 799-821.

[a] Acknowledgment of sources is an appropriate academic norm, but their express permission is not required.

4

Assessment and Recommendations

Our nation has a vibrant and demonstrably productive community of scientific and technical (S&T) database creators, disseminators, and users that has led the world. Advances in computing and communication technologies make S&T databases and the facts they contain increasingly valuable for producing new discoveries and for accelerating the growth of knowledge and the pace of innovation. The same technologies that facilitate the effective production, dissemination, and use of data, however, can also expedite their unauthorized dissemination and use, with the potential effect of undermining incentives to create new databases, facilitating unfair competition and wholesale piracy, and in the most extreme cases, exposing the original database rights holder to market failure.

As Chapter 3 points out, the current efforts to enact statutory federal database protection in the United States appear to be stimulated by three principal factors: (1) the possibility for rapid and complete database copying with the potential for instantaneous broad dissemination; (2) the gap in U.S. law created by the *Feist*[1] decision, which invalidated copyright protection on the basis of investment and effort (i.e., "sweat-of-the-brow") investments alone; and (3) the E.U. Database Directive, which requires other nations to pass a similar law in order for their citizens to enjoy the E.U. Directive's protections in Europe, thereby providing a potentially unfair advantage to European competitors of the U.S. private sector.[2]

---

[1] *Feist Publications, Inc. v. Rural Telephone Service Co.*, 499 U.S. 340 (1991).

[2] Although certain database vendors might be at some competitive disadvantage in the European Union, the committee believes that a less protectionist law in the United States that encourages the

73

The committee believes, however, that the need for additional statutory protection has not been sufficiently substantiated. The high level of activity in the production and use of digital S&T (and other) databases in the United States serves as prima facie evidence that threats of misappropriation do not constitute a crisis. Nor do the existing legal, technical, and market-based measures provide a chronic state of underprotection for proprietary databases. The almost universal use of licensing, rather than sale, of online databases and other digital information, coupled with technological enforcement measures, on balance potentially provides much stronger protections to the licensors vis-à-vis their customers than they enjoyed prior to *Feist* and under the print media copyright regime (see Table 3.2 in Chapter 3). While some of the current law providing protection to database rights holders remains uncertain in terms of scope of applicability, the trend in recent years has been to broaden, rather than narrow, applicable intellectual property protections.

Moreover, strong statutory protection of databases would have significant negative impacts on access to and use of S&T databases for not-for-profit research and other public-interest uses. Nevertheless, although the committee opposes the creation of any strong new rights in compilations of factual information, it recognizes that limited new federal legal protection against wholesale misappropriation of databases may be appropriate. In particular, a balanced alternative to the highly protectionistic E.U. Database Directive could be achieved in a properly scoped and focused new U.S. law, one that might serve as a model for an international treaty in this area.

In this chapter, the committee examines several legislative options and related government activities, and recommends a number of legislative principles and policy actions to help inform the current debate. The chapter concludes with a recommendation directed specifically to the not-for-profit S&T community.

ASSESSMENT OF LEGISLATIVE OPTIONS, WITH RECOMMENDATIONS ON GUIDING PRINCIPLES

The committee assessed and compared three separate proposals for increased database protection of private-sector databases in the United States that were placed in the *Congressional Record* at the beginning of the 106th Congress.[3] During the 105th Congress, the House of Representatives twice approved a measure establishing a specific statutory scheme for the protection of databases— H.R. 2652[4] and Title V of H.R. 2281,[5] which was substantially the same as H.R.

---

use of factual data for both public interest and commercial purposes will benefit the U.S. economy and society to a greater extent.

[3] *Cong. Rec.,* Vol. 106, S. 316 (Jan. 19, 1999).

[4] H.R. 2652, the "Collections of Information Antipiracy Act," 105th Congress (1997).

[5] H.R. 2281, Title V, the "Collections of Information Antipiracy Act," 105th Congress (1998).

2652.[6]   Both H.R. 2652 and 2281 were adopted on suspension calendar by the House despite significant opposition from an array of scientific, educational, library, and consumer public-interest organizations and institutions, as well as from a number of commercial publishers and information technology services companies.[7]   These House bills applied both to databases that qualified for copyright protection and to noncopyrightable "sweat-of-the-brow" databases that did not.

The proposed legislation drew the concerned attention of the not-for-profit communities because expanding private property rights in factual databases could interfere with scientific progress and other public-interest uses of data.[8]   At the same time, some private-sector firms believe that their databases are vulnerable to misappropriation due to a gap in the law.[9]   Other reactions from the private sector included the perception that the proposal adopted by the House placed too many impediments to transformative uses of existing databases for commercial

---

[6] The only significant change in Title V of H.R. 2281 was to remove "potential markets" from the ambit of liability for not-for-profit uses in Section 1403, Permitted Acts (a) Educational, Scientific, Research, and Additional Reasonable Uses, which was amended as follows:

(1) Certain Not-for-profit Educational, Scientific, or Research Uses.— . . . no person shall be restricted from extracting and using information for not-for-profit educational, scientific, or research purposes in a manner that does not harm directly the actual [or potential] market for the product or service referred to in section 1402." [words in brackets deleted].

[7] See the testimony of the not-for-profit sector cited in note 8 below, and of the commercial opponents to the legislation in note 10 below. In addition, over 130 organizations and companies signed a position statement critical of H.R. 354 that was placed in the public record by Dr. James Neal, director of the Milton S. Eisenhower Library at Johns Hopkins University and president of the Association of Research Libraries, during the March 18, 1999, Hearing on H.R 354, the "Collections of Information Antipiracy Act," held by the Subcommittee on Courts and Intellectual Property of the Committee on the Judiciary of the U.S. House of Representatives [hereinafter March 18, 1999, Hearing]. A copy of the position statement and the full list of signatories may be found online at <www.databasedata.org>.

[8] See testimony by Wm. A. Wulf, president of the National Academy of Engineering on behalf of the National Academies, J.H. Reichman, professor at the Vanderbilt University School of Law, and James G. Neal, director of the Milton S. Eisenhower Library at Johns Hopkins University and president of the Association of Research Libraries, at the October 23, 1997, Hearing on H.R. 2652, the "Collections of Information Antipiracy Act," held by the Subcommittee on Courts and Intellectual Property of the Committee on the Judiciary of the U.S. House of Representatives [hereinafter October 23, 1997, Hearing].

[9] Id. See testimony by Paul Warren of Warren Publishing, Inc., on behalf of the Coalition Against Database Piracy. See also testimony by Robert E. Aber, senior vice president and general counsel, the NASDAQ Stock Market, Inc., on behalf of the Information Industry Association, at the February 12, 1998, Hearing on H.R. 2652, the "Collections of Information Antipiracy Act," held by the Subcommittee on Courts and Intellectual Property of the Committee on the Judiciary of the U.S. House of Representatives [hereinafter February 12, 1998, Hearing].

purposes.[10]   Perhaps most significant, the Administration presented its own consensus critique of the House bills on August 4, 1998,[11] and the Department of Justice[12] and the Federal Trade Commission[13] issued legal memoranda outlining their concerns about the legislation's constitutionality and anticompetitive effects, respectively.

Soon after the approval of Title V of H.R. 2281 by the House in July 1998, Senator Orrin Hatch (R-UT), chairman of the Senate Committee on the Judiciary, initiated negotiations among the various interests, which continued from early August until early October.[14]   While substantial progress was made in this process and the needs of the science, education, and library communities were directly acknowledged, a consensus was not achieved before the 105th Congress adjourned in October 1998.[15]

Shortly after the 106th Congress convened in January 1999, Congressman Howard Coble (R-NC), chairman of the Subcommittee on Courts and Intellectual Property of the House Committee on the Judiciary, reintroduced as H.R. 354[16] the proposal that had twice passed the House in the previous session.   H.R. 354 included two changes to respond to concerns of the scientific community and other critics of the original legislative proposal.[17]   Thereafter, Senator Hatch

---

[10] Id. See testimony by Jonathan Band, partner, Morrison & Foerster LLP, on behalf of the On-Line Banking Association, and by Tim Casey of MCI, Inc. on behalf of the Information Technology Association of America at the February 12, 1998, Hearing.

[11] See letter from Andrew J. Pincus, general counsel of the Department of Commerce, to The Honorable Orrin G. Hatch, chairman, Senate Committee on the Judiciary, August 4, 1998, summarizing "a number of concerns" of the Administration with H.R. 2652.

[12] See memorandum for William P. Marshall, associate White House counsel, from William Michael Treanor, deputy assistant attorney general, Office of Legal Counsel, Department of Justice, July 28, 1998, regarding "Constitutional Concerns Raised by the Collections of Information Antipiracy act, H.R. 2652."

[13] See letter from Robert Pitofsky, chairman, Federal Trade Commission, to The Honorable Tom Bliley, chairman, Committee on Commerce, U.S. House of Representatives, September 28, 1998, regarding potential anti-competitive effects of Title V of H.R. 2281.

[14] These negotiations were conducted in closed sessions with representatives of the principal organizations that had previously taken a public position on the House bills. The Intellectual Property Counsel to Senator Hatch, Edward Damich, moderated the negotiation process.

[15] For a detailed discussion of the Senate negotiations and the legislative process associated with the database protection legislation in the U.S. Congress through early April 1999, see generally J.H. Reichman and Paul F. Uhlir (1999), "Database Protection at the Crossroads: Recent Developments and Their Impact on Science and Technology," *Berkeley Technology Law Journal*, Vol. 14, pp. 793-834.

[16] H.R. 354, the "Collections of Information Antipiracy Act," 106th Congress (1999).

[17] The two changes made in H.R. 354 by the House Subcommittee on Courts and Intellectual Property included an attempt to eliminate the potential for indefinitely prolonging the 15-year duration of protection in section 1408 (c), and expanding the scope of the exemption for certain not-for-profit educational, scientific, and research uses in section 1403 (a), both of which are discussed in more detail later in this chapter.

inserted in the *Congressional Record* two other proposals—the first drafted by a coalition of commercial and not-for-profit interests (hereinafter, the Coalition Proposal)[18] seeking much more limited protection than H.R. 354, and the second a draft bill that had emerged at the end of the 1998 negotiations sponsored by Senator Hatch (hereinafter, the Senate Discussion Draft).[19]

In the rest of this section the committee discusses several of the most important provisions of these three proposals and evaluates them in terms of their potential effects on access to and use of S&T data by public-interest users. In doing so, the committee recognizes that these proposals have changed and will continue to change before any one of them is considered for final adoption. Nonetheless, they serve as models for the types of issues that arise from the perspective of the research and education communities confronted with the prospect of legislative changes that would affect access to data.[20]

As noted in Chapter 1, because of the complex web of interdependent relationships among public-sector and private-sector database producers, disseminators, and users (see Table 1.1 in Chapter 1 for an indication), any action to increase the rights of persons in one category is likely to compromise the rights of the persons in the other categories, with potentially far-reaching negative consequences. A principal concern of the committee, therefore, is that the development of any new database protection measures aimed at protecting private-sector investments take into account the need to promote access to and subsequent use of S&T data and databases not only by the not-for-profit sector, but by commercial producers of derivative databases as well. Of course, it is in the common interest of both database rights holders and users—and of society generally—to achieve a workable balance among the respective interests so that all legitimate rights remain reasonably protected. Therefore, **as a general guiding principle, the committee recommends that any new federal protection of databases should balance the costs and benefits of the proposed changes for both database rights holders and users.**

---

[18] See the "Database Fair Competition and Research Promotion Act of 1999," *Cong. Rec.*, Vol. 106, S. 316 (Jan. 19, 1999).

[19] Id., "Chapter 14—Protection of Databases," S. 322-326.

[20] At the time of this writing, the House Committee on Commerce has introduced and marked up a slightly modified version of the Coalition Proposal. See H.R. 1858, The Consumer and Investor Access to Information Act of 1999, 106th Congress, May 20, 1999. The House Committee on the Judiciary also has marked up H.R. 354, which includes a number of significant revisions. Because the study committee had already written its report, it was not able to consider these additional changes to the proposed legislation. Nevertheless, the committee believes that its analysis and recommendations remain relevant to the ongoing debate concerning this legislation, as well as to any eventual implementation of a statutory database protection regime. Any bill that is finally adopted, if any, most likely will be substantially further modified. For this reason, the committee presents its legislative recommendations as guiding principles, rather than as specific legislative language.

The Standard of Harm

The key provision of the three legislative proposals defines the nature of protection accorded a database rights holder and establishes the standard of harm against which a defendant's liability is to be judged. As introduced in January 1999, H.R. 354 prohibited the "extraction or use" of a substantial part of a database if it results in "harm to the actual or potential market" for any product or service incorporating the database.[21] A "potential market" includes any market the database rights holder "has current and demonstrable plans to exploit" or a market that is "commonly exploited by persons offering similar products or services." The Senate Discussion Draft narrowed the protection of actual markets to those markets commonly exploited by persons offering similar products.[22]

The Coalition Proposal took a different approach, prohibiting only the "duplication of another's database [and inclusion of those records in] . . . a database that competes with the original."[23] To compete with the original database, the duplicate must be substantially identical to the original, must be shown to displace substantial sales or licenses of the original, and must be offered for sale or digitally distributed in such a manner as to "significantly diminish the incentive to invest" in developing the original database. The latter requirement may be interpreted as threatening the opportunity to recover a reasonable return on the investment in collecting or organizing the original database.

None of the three legislative proposals purported to create broad property rights in the original database, as the E.U. Directive does. However, by expanding protection to "potential markets," H.R. 354 would allow the rights holder to foreclose markets or uses beyond the rights holder's actual use. This has the effect of granting exclusive rights to the original database rights holder in uses unknown at the time of the database's creation. The limitation of the H.R. 354 language stating "current demonstrable plans to exploit" is unclear because the time at which "current plans" is to be measured is not stated. Does "current" mean at the time the extraction and use occur, at the time the user develops the new market, or at the time the database rights holder brings suit? A scientific researcher might discover an entirely new application for a database, only to be foreclosed from such use if the original database rights holder were subsequently to develop "current and demonstrable plans" to exploit that application as an additional market. For example, scientist A has a database consisting of human gene sequences potentially useful for locating genes controlling certain diseases, but does not know of any particular sequences that are valuable for this purpose. By extraction and use of A's database, scientist B discovers a set of sequences

---

[21] H.R. 354, section 1402.

[22] Section 1301(3).

[23] Section 1401. The House Committee on Commerce bill, H.R. 1858, has extended that prohibition to include a "discrete section" of a database.

that seem particularly valuable for further experimentation and makes this subset of sequences available to the scientific community. In doing so, scientist B could violate the protection provided by H.R. 354. Although protection of original, noncopyrightable databases with a strong, copyright-like property right may encourage additional investment in producing databases, it simultaneously discourages others from investing in discovery of new uses for existing databases and elevates the cost of using them. In principle, the public benefits most from the weakest legal incentives for encouraging such investments, and intellectual property theory has always promoted the open availability of facts. For the creation of legal incentives greater than this, the former chairman of the House Committee on the Judiciary, Robert Kastenmeier, required proponents of new intellectual property rights to meet a very heavy burden.[24]

The Senate Discussion Draft provided considerably narrower protection by requiring a showing of "substantial" harm to the actual or neighboring market,[25] which was defined in the Proposed Conference Report Language as "harm [that] is such as to significantly diminish the incentive to invest in gathering, organizing or maintaining the database."[26] The harm test in the Coalition Proposal was similarly circumscribed, requiring both a displacement of substantial sales and a showing that the unauthorized use "significantly diminished the incentive to invest in the collecting or organizing of the protected database.[27] These latter two formulations expressly acknowledged that not all duplications are actionable, even if used for commercial purposes (e.g., in distant markets) or for procompetitive purposes by honest means. The intent was to recognize that competitors who add value and generate socioeconomic benefits should not incur liability if they do not directly harm the market of the original database rights holder, i.e., if they do not compete unfairly.

The committee believes that strong protection based on a broadly framed standard of harm test, such as the one proposed by H.R. 354, poses a number of potential problems for research, education, and other public-interest users, as well as for legitimate private-sector, value-adding database producers. As a general rule, the stronger the statutory protection, the greater the encumbrances will be on the reuse and transformation of data received by second-generation database producers and users. One person's derivative use can be characterized as an infringement on the original database rights holder's product; where the bar is set will determine to what extent database producers and disseminators will be

---

[24] See Robert W. Kastenmeier and Michael J. Remington (1985), "The Semiconductor Chip Protection Act of 1984: A Swamp or Firm Ground?," *Minn. L. Rev.*, Vol. 70, p. 417, establishing a stringent four-part test for assessing the merits of any proposed intellectual property protection for new technologies.

[25] Section 1302.

[26] Proposed Conference Report Language, Section 1302, at 33.

[27] Section 1405 (4).

enriched at the expense of all socially and economically valuable downstream uses.

As noted in Chapter 1 (and indicated in Table 1.1), many organizations are users as well as producers and vendors of S&T databases, as, for example, when they draw on one or more databases to search for cross-disciplinary associations or to create a derivative or value-added database targeted to a competing or entirely new market. Private-sector creators of derivative databases have conflicting views of protection: protection of source databases might deprive them of access, but insufficient protection for their own creations might make them vulnerable to copying. Protection entails contradictory consequences for creators of derivative databases. **A concern of the committee, therefore, is that any new protection judged to be necessary must take into account the need to promote access to and subsequent use of S&T data and databases not only by the not-for-profit sector, but by commercial producers of derivative databases as well.**[28]

A major negative effect of a strong standard of harm test would be to raise the resale prices for value-added or derivative databases, as well as to inhibit their production. Value-adding database producers that use multiple data sources to create new products, as is common in both the private and the public sector, are particularly penalized by a strong standard of harm test.[29] Although the consequences would be difficult to measure, strong new rights for database rights holders would probably result in a broad loss of research opportunities.[30] If, for example, potential users opted to engage in other professional activities rather than deal with more expensive and onerous restrictions on database use, the probability of subsequent discoveries, innovations, and advances in knowledge would decrease, not only because of the reduced number of users, but also because the remaining database users would be constrained in their activities. Downstream commercial providers who must pay license fees to the rights holders of sole-source databases can recover such fees only if they themselves charge more for access, costs that are passed down the chain of derivative products to all users, including investigators in not-for-profit institutions.

By making entry into a market more expensive, greater statutory protection also could increase the likelihood that small or niche markets, which are com-

---

[28] Indeed, many commercial entities have expressed concerns about such effects of strong database protection. See, for example, the testimony and position statement cited in note 10.

[29] As noted by Nobel laureate Joshua Lederberg in his testimony on behalf of the National Academies and the American Association for the Advancement of Science at the March 18, 1999, Hearing, note 7, the "recent advent of digital technologies for collecting, processing, storing, and transmitting data has led to an exponential increase in the number of databases created and used. A hallmark trait of modern research is to obtain and use dozens, or even hundreds of databases, extracting and merging portions of each to create new databases and new sources of knowledge and innovation."

[30] See Reichman and Uhlir (1999), note 15, p. 820.

monplace for many S&T databases, would be served by sole-source providers. A higher cost of entry typically deters entrants and allows the first entrant to act as a monopolist.[31] A sole source may then use its market power to inhibit the development of derivative databases if these are interpreted as undermining the investments in the original database, even if such derivative uses are in completely different markets or are protected as "permitted acts" under a statute. Monopoly power could be exercised over the data in many areas of research, because most observational databases cannot be independently recreated after the fact, and it is economically inefficient and undesirable to require independent, redundant collection of original data in activities that use very high cost systems. As the Federal Trade Commission cautioned in its analysis of the predecessor bill to H.R. 354, ". . . policies that further entrench the market power of single-source data providers could have an unintended, undesirable impact on competition and innovation because of the significant potential for anticompetitive conduct in single-source database markets."[32] The law should encourage competition because competition leads to lower prices, resulting in broader use and, hence, further discovery and innovation.

Increased license fees or unreasonable restrictions on subsequent uses or redissemination of data would negatively affect both government and not-for-profit database value adders or disseminators in other ways as well. For example, European government meteorological data providers, who are already benefiting from the stronger protections offered by the E.U. Database Directive, are placing various use and redistribution restrictions on the National Oceanic and Atmospheric Administration (NOAA), asking NOAA to enforce these restrictions in the United States, contrary to existing U.S. law and policy. Such encumbrances from private-sector sources would be exacerbated by any database legislation that, similar to the E.U. Database Directive, extended protection to elements not now protected. Government S&T managers, in particular, are concerned that they do not receive enough funding to pay license fees and enforce restrictive provisions, in addition to meeting the costs of data collection and database preparation, and anticipate that they might have to decline data contributed by private-sector sources (as well as public-sector European sources) that carry high royalties or restrictions on subsequent distribution that require enforcement by the user.

---

[31] See Laura D'Andrea Tyson and Edward F. Sherry (1997), *Statutory Protection for Databases: Economic & Public Policy Issues*, research paper prepared under contract to Reed-Elsevier, Inc. and The Thomson Corporation, and presented as testimony on behalf of the Information Industry Association at the October 23, 1997, Hearing, note 8. Tyson and Sherry, however, generally argue that there are not many instances in the commercial database industry in which sole sources dominate the market and can prevent or inhibit entry. Although the committee did not analyze the entire database market in this study, it did find that in many S&T areas, including practically all observational databases, the data sources are unique.

[32] Federal Trade Commission letter, note 13, p. 2.

With increased statutory protection for databases and the accompanying higher transaction costs, scientists and educators in the not-for-profit sector might no longer be able to afford access to newly proprietary data sources or to enforce subsequent access and use restrictions on the data obtained from those sources, contrary to existing norms and practice.[33] Not-for-profit research institutions tend to be conservative, risk-averse organizations that err on the side of caution, and they would likely institute guidelines prohibiting any database research activity that might potentially expose them to liability under a new legislative regime and to costly litigation. Such a possibility is particularly problematic given the uncertainty about what portions of databases would be deemed "qualitatively substantial" by the rights holders in each case and about what they would view as a "reasonable use" by not-for-profit entities. Such defensive measures would serve to further restrict, perhaps even beyond what might be allowed under the law, what scientists and educators can do with databases that they lawfully obtain.

Providing stronger property rights for databases that contain information of high commercial value, such as in the area of genomic research, can have the opposite of the intended effect, because the price of access to these databases is inversely related to the number of users who will access them. Hence, from an S&T perspective, the goal is to encourage the generation of publicly funded, and therefore readily available, collections of data in key scientific areas, where the use of this information is of potentially great commercial value, and to discourage the tendency for private companies to capture this information and restrict access to a limited audience. Promoting broad access to publicly generated databases has the additional benefit of fostering active competition and value-adding activity since all commercial and academic organizations would have access to this information.

Moreover, enhancing database protection would also serve as an incentive to both government agencies and not-for-profit organizations to privatize or commercialize their research databases. Such action would have the undesirable outcome of reducing the number of databases in the public domain and thus would have a chilling effect on the full and open data exchange and sharing ethos that benefits so many areas of scientific and engineering research.

Since a strong case for significantly greater protection of databases has not been made, primarily because existing protections already go a long way toward protecting database providers, the committee believes that the concerns regarding

---

[33] Examples of this problem are already abundant in the restrictions on experimental research uses of patentable or otherwise protected innovations in the biotechnology sector. See M.A. Heller and R.S. Eisenberg (1998), "Can Patents Deter Innovation? The Anticommons in Biomedical Research," *Science*, Vol. 280, p. 698. Such a problem would be more insidious in the case of noncopyrightable factual databases, which are subpatentable innovations that do not merit strong property rights and that have been used much more widely and openly in research to date.

increased encumbrances on access and use, as well as the potential for higher prices and related transaction costs, cannot be ignored. **In light of these concerns, the committee recommends that any new federal statutory protection of databases should limit any additional protection to prohibition of acts of unauthorized taking that cause substantial competitive injury to the database rights holder in the rights holder's actual market. The standard of harm should be sufficiently clear to permit good-faith users to know when they are infringing on a database rights holder's rights and should not undermine the nation's capabilities for innovation or competition in the marketplace.** Such a formulation would help prevent undue and inappropriate interference with scientific inquiry and with other traditional and customary public-interest uses of data, as well as promote legitimate and socially beneficial commercial competitive activities.

Scope of Protection

The first section of all three of the legislative proposals considered by the committee defined a database as a "collection of information collected and organized for the purpose of facilitating access to discrete items of the information." All three proposals also provided protection to databases developed through the investment of substantial monetary or other resources. "Information" was defined to include data, facts, or other intangible material capable of being collected and organized in a systematic way. The Coalition Proposal, however, excluded "works of authorship"—a term applicable to subject matter protected by the copyright system. Such an exclusion would deny protection to copyrightable works such as anthologies of an author's works or a scientific journal that might otherwise be regarded as a database of articles. The H.R. 354 and Senate Discussion Draft proposals included these works, consistent with the subject-matter scope of the E.U. Directive. The committee believes that the inclusion of collections of works of authorship, which are already unambiguously protected by copyright, is both unnecessary and unsupportable. If the purpose of this legislation is to fill a purported gap in the legal protection currently available to noncopyrightable databases, then that scope of protection should not extend so broadly as to cover fully copyrightable anthologies, journals, and textbooks. **The committee therefore recommends that the subject-matter scope of any new federal statutory protection of databases be constrained to databases comprising a collection of discrete facts and items of information, and expressly exclude collections of copyrightable material, which is already protected. Further, protection under any new statute should extend only to a database that is the product of a substantial investment, and not to any idea, fact, procedure, system, method of operation, concept, principle, or discovery disclosed by the database.**

Term of Protection

H.R. 354[34] and its predecessor bills, as well as the Senate Discussion Draft,[35] provided protection for a nominal term of 15 years, reflecting an effort to meet the E.U. Database Directive's term of protection[36] in response to the E.U. Directive's attempt to assert a reciprocity requirement. However, no analysis or empirical study supports the choice of 15 years of protection by the European Union, in comparison with other potential terms of protection. The committee believes that unquestioningly adopting an arbitrary term of protection developed by a foreign power without any experience as to its potential effects would be unsupportable. Further, the European Union's reciprocity requirement contravenes a well-established U.S. government policy requiring national treatment under all international intellectual property law agreements.[37] Nonetheless, one key difference between the two congressional proposals above and the E.U. Directive is that the E.U. Directive allows the period of protection for the entire database to be extended for another 15 years with each substantial new investment—thereby providing the rights holder with the possibility of perpetual protection of the entire database—while the two congressional proposals tried to limit the extension of protection to the new data that might be added.[38]

Despite these efforts to limit protection to 15 years, the committee is concerned about such a long term of initial protection for factual data, whether for research and education purposes or any other uses, especially if the standard of harm remains as strong as proposed in H.R. 354. Indeed, there has been a complete failure on the part of the proponents for a 15-year term of protection either to justify that term, independent of the arbitrary decision made by the framers of the E.U. Directive, or to compare it with other, shorter terms of

---

[34] Section 1408(c).

[35] Section 1310(c).

[36] Article 10.

[37] See Andrew Pincus testimony on behalf of the Administration from March 18, 1999, Hearing, note 7, which states in part: "The Administration opposes such 'reciprocity' requirements, both domestically and internationally. We believe that commercial laws (including intellectual property and unfair business laws) should be administered on national treatment terms, that is, a country's domestic laws should treat a foreign national like one of the country's citizens. This principle is embodied in Article 3 of the Agreement on Trade-Related Aspects of Intellectual Property Rights (TRIPS Agreement) as well as more generally in the Paris Convention for the Protection of Industrial Property and the Berne Convention for the Protection of Literary and Artistic Works. The Administration believes that Congress should craft U.S. database protection to meet the needs of the American economy. . . .", p. 32.

[38] Failure to limit the term of protection, coupled with a strong proprietary right such as the one proposed in H.R. 354, has led some legal commentators to question the constitutionality of such a provision under U.S. law. See generally William Michael Treanor (1998), note 12, and Marci Hamilton, Cardozo Law School, letter to Howard Coble, chairman of the Subcommittee on Courts and Intellectual Property, House Committee on the Judiciary, February 10, 1998, 5 p.

protection used in other intellectual property law models.[39] Although, in comparison, the term of copyright is much longer, historically the term has been justified and set according to an author's likely lifetime, plus some additional time in which the author's heirs can benefit.[40] Moreover, there are other constitutionally imposed limits on both the scope and length of such protection. If a database meets the constitutional and statutory requirements for copyright, then the rights holder can obtain the longer term of protection that copyright law affords. However, the committee has been unable to find any rationale for the 15-year term and, based on market factors relevant to databases, questions that length of protection for noncopyrightable databases or "substantial portions thereof."

The committee notes that the average high-activity life span of original data in an online commercial database is approximately 3 years.[41] Consequently, most of the incentive for creating and distributing databases comes from the return on investment achieved in the first 3 years, when demand for and use of databases are highest. It is important to note, however, that there is a significant difference between how long databases have value and how long statutory protection for noncopyrightable databases should be accorded. Although there are S&T databases in rapidly progressing areas of research, such as some in the life sciences, whose research and commercial values plummet very quickly as new data supersede old, many research endeavors, such as the study of environmental trends, longitudinal socioeconomic studies, and various types of historical analyses, not only depend on consistently collected long-term data sets, but also require access to both current and historical data for comprehensive and comparative study and verification. The committee believes that any term of protection that is set should have a duration deemed sufficient to create incentives for producing original new databases. It should not be set to assure that rights holders capture all the value, since that would require an exceedingly long period to cover all cases and would constitute establishing protected markets that are inappropriate. As a general rule, the broader and stronger the scope of protection, the shorter the period of protection needs to be to provide an appropriate incentive for creating a database. In any event, the case for the term of protection to be used should be made by those who are seeking the new protection, and this has not been done.

Neither H.R. 354[42] nor the Senate Discussion Draft[43] completely addressed

---

[39] See, for example, J.H. Reichman and Pamela Samuelson (1997), "Intellectual Property Rights in Data?" *Vanderbilt Law Review*, Vol. 50, pp. 137-152.

[40] See 17 U.S.C., section 302.

[41] See generally, Martha E. Williams (1984-1999), *Information Market Indicators: Information Center/Library Market—Reports 1-60*, Information Market Indicators, Inc., Monticello, IL.

[42] Section 1408(c).

[43] Section 1310. However, the Hatch Discussion Draft did include a provision for voluntary deposit of databases to the Copyright Office in Section 1311.

the potential problem of extending protection to substantial portions of old data that may become inseparably intermingled with new data, which raises the issues of how to effectively track such activity and how to provide adequate notice of which substantial portions remain subject to statutory protection.[44] Important factors to consider in this regard are scientific and legal authentication methods and the necessary documentation (metadata) requirements. Adequate database authentication and documentation are essential not only as prerequisites for accurately tracking the term of protection for any given database, but also for improving the reliability and value of databases for research and other uses. In addition, although the committee did not explicitly consider it, some observers have noted that a registration system similar to the one administered by the Copyright Office for copyrighted works could be helpful in notifying users about the expiration of the term of protection for any given database.[45]

Both proposals also allowed for 15 years of retroactive application of protection by not limiting causes of action to databases created on or after the date of enactment of the legislation. The committee finds little justification for legislation that is supposed to be necessary to stimulate and protect new investment to apply to databases already created without the benefit of such protection.

The Coalition Proposal traded off a much weaker standard of harm and scope of protection for an unlimited term of duration—basically for as long as the database has some commercial value to the rights holder. The cause of action was limited to a duplication of a database that is placed in direct commercial competition,[46] and this was coupled with additional exemptions for scientific, educational, or research uses.[47] The committee believes that the unlimited term of duration proposed in the Coalition Proposal is not unreasonable in the overall context of that proposed bill, since it would also provide some added protection to those databases that have significant commercial value beyond 15 years and that have long existed. It is important to emphasize, however, that any further strengthening of the standard of harm under this proposal would likely make the unlimited term of duration not only unacceptable, but unconstitutional.[48] Finally, neither the problem of identifying substantial portions of old versus new data nor the problem of retroactive application of the statute arose in the Coalition Proposal.

As a general principle, the committee recommends that the term of protection in any new federal statutory protection of databases be limited to a period of time sufficient to provide incentives found necessary for the

---

[44] For a discussion of how such legislation might address this problem, see testimony of Andrew J. Pincus, March 18, 1999, note 7.

[45] Id.

[46] Section 1405(4).

[47] Section 1402(e).

[48] See documents referenced in note 38.

creation of new databases. If legislation with a fixed term of protection is adopted, an appropriate term of protection most likely should be substantially shorter than the proposed 15-year term. It should also be based on an analysis of the economics of the database industry, rather than set arbitrarily.

The committee also recommends that any new legislation with a fixed term of protection also should require database rights holders to provide notice of expiration of the term of protection. Specifically, any such legislation should:

• Require database rights holders to identify the date on which the database was created so that the user will know when it no longer enjoys statutory protection (of course, those databases that remain commercially valuable longer than the statutory period of protection can continue to be protected by other means, such as copyright, trade secret, contract, and technical and other measures); and

• For databases that are updated continuously, or at periodic intervals, require database rights holders to identify with reasonable precision those substantial portions of the database that are and are not subject to protection. Failure to identify the date of creation for each new substantial portion of a database should serve as a basis for a defense against infringement after the expiration of the term of protection for the original portion of the database.

Finally, the committee recommends that protection be applied only to databases created after the effective date of any new legislation, in recognition that a major purpose of enacting enhanced protection is to provide additional incentives for the development of new databases.

Exemptions for Not-for-Profit Research and Education

It might be asked why not-for-profit research and education should have access to data on better terms than commercial enterprises, and why those communities should get special "subsidies" from database producers. After all, they do not receive parallel subsidies from suppliers of laboratory mice or telescopes.

In Chapter 2, the committee argues that the price of access to commercial databases will typically be higher than the efficient-access price, leading to inefficient use. (This is not true of (nonproprietary) laboratory mice or telescopes. Mice and telescopes do not have the increasing-returns cost structure of a database, and their price in a competitive market will be the "efficient-access" price.) The consequent reduction in data use could negatively affect both commercial users and public users of databases, but commercial users have an advantage. They can recover some of the high access fees by pricing their own products appropriately. In contrast, the revenues of public users, such as public research laboratories and universities, come mostly from public agencies (taxpayers). The scientific community legitimately predicts that any increase in the price of access to data will not be compensated by increased public subsidies.[49] Hence publicly funded users are likely to be more negatively affected than commercial users by new database protection legislation.

The consequent reduction in resources available for education and science would be particularly damaging because education and science have a public-good aspect. They generate "nonexcludable benefits," sometimes called externalities, that go beyond any benefits that could be realized on their own ledgers or to their own constituents. All of society benefits when children are educated, when a cure is discovered for a disease, or when significant trends in climate are detected, and society will receive these benefits even without paying directly for them. To ensure an appropriate level of investment in these activities—a level proportionate to the benefits likely to be achieved—some public subsidy is required. This explains, in part, the long-standing U.S. tradition of public education and public funding of research—a tradition that can claim a significant share of the acclaim for our economic and social standing.

The privileged status of public-interest data users was in fact recognized to varying degrees by all three of the legislative proposals under consideration. They all attempted to grant education, science, and research more leeway in utilizing protected databases, but they varied with respect to their degree of anticipated effectiveness.

H.R. 354 and the Senate Discussion Draft initially permitted extraction or use of information for not-for-profit educational, scientific, or research purposes, as long as the use does not interfere with the database rights holder's "actual market."[50] Under this provision, research that produces a product that potentially, or in fact, opens a new market not exploited by the database rights holder would not violate the law. Both proposals went further, to protect some educational or research activities, even if they do interfere with the rights holder's actual market. H.R. 354 provided a fact-dependent exemption similar to the fair-use privilege in copyright law.[51] An individual act of use or extraction of another's database for teaching or research would be privileged if reasonable under the circumstances.[52] The reasonableness of the use or extraction was to be determined by consideration of four factors: the commercial or not-for-profit nature of the use or extraction, the good faith of the user, whether the portion used or extracted is incorporated into an independent work, and whether the use or extraction is in the same field as the original database. However, notwithstanding

[49] See National Research Council (1997), *Bits of Power: Issues in Global Access to Scientific Data*, National Academy Press, Washington, D.C., p. 114.

[50] Section 1403(a)(1) in H.R. 354, and section 1303(c) in the Hatch Discussion Draft.

[51] See 17 U.S.C., section 107, and the discussion of fair use in Chapter 3.

[52] Section 1403(a)(2).

these factors, a use or extraction would not be privileged if it was likely to become a market substitute for all or part of the original database.[53]

The Senate Discussion Draft proposed that anyone can use a protected database for purposes of "illustration, explanation or example, comment or criticism, internal verification, or scientific or statistical analysis of the portion used" and further authorized not-for-profit scientific, educational, or research activities "for similar customary or transformative purposes."[54] This exemption was limited if substantial harm accrues to the original database rights holder because the use was more than reasonable and customary, consists of a substitute for the original database, is intended to avoid payment of reasonable fees for use of a database specifically marketed for education, scientific, or research purposes, or is a part of a pattern of systematic use.[55]

Because the Coalition Proposal prohibited the duplication of a database only if the duplication displaces substantial sales or licenses of the original database, there was less need for a strong research privilege. Nonetheless, that proposal provided that uses for science, research, or education are privileged unless the uses are part of a "consistent pattern" designed to compete directly with the original database, or to avoid reasonable fees for access to a database specifically designed for a scientific, research, or educational program.[56] Under the Coalition Proposal, the use of another's database for scientific or technical research would be permitted unless the use directly undermines the incentive to invest in the original database.[57] This model, based on unfair competition law, strongly acknowledged the value of promoting unfettered scientific research but still offered protection to prevent database rights holders from being the victims of commercial misappropriation. Finally, the Coalition Proposal recognized the fairness of reasonable access charges for databases whose only purpose is for scientific research.[58]

However, H.R. 354 and, to a lesser extent, the Senate Discussion Draft would represent a considerable risk for the conduct of research and education. The nature and value of the products of research frequently are unknown until well after the research is conducted. Scientists will not know whether a particular use of a database under H.R. 354 will affect the "actual market" for the original or whether a court will subsequently weigh the factors in such a way as to make the use privileged. The Senate Discussion Draft, which allowed not-for-profit scientific research for "similar customary or transformative purposes," still withdrew the privilege if the ultimate result would be "likely to serve as a substitute" for the

---

[53] Section 1403(a)(2)(A).
[54] Section 1304(a).
[55] Section 1304(b).
[56] Section 1402(e).
[57] Section 1405(4).
[58] Section 1402(e).

original database.[59]   Thus, under both of these proposals, a researcher could not be certain of the authority to utilize another's database until the results of the research were known. Also, neither a researcher nor any other user would know with any certainty what the database rights holder might consider to be a quantitatively or qualitatively substantial part of the rights holder's database, and the rights holder would have every incentive to define its protected domain as broadly as possible. This uncertainty could discourage uses that would otherwise occur under a more privileged access and use provision.[60]   Moreover, the "actual market" of the originator might be undermined in many ways other than by direct competition, such as by research demonstrating the original database was inaccurate, was built on false premises, or did not do what it was marketed to do.

The uncertain results of research make license transactions for the use of databases intrinsically difficult. Although a licensor will want to establish a fee that protects against any reduction in value of the licensor's database and that provides for sharing in the economic value of the resulting product, not knowing the value of the research results in advance confounds defining "the harm to the original database" or the database value.

Both H.R. 354 and the Senate Discussion Draft also prevented patterns of systematic extraction or uses of individual items of information or other insubstantial parts of a database,[61] and the H.R. 354 exemption for science was limited to an "individual act of use."[62]   Both of these limitations failed to reflect the nature of some scientific research. For example, the potential for new treatments for disease represented by the development of databases arising out of the Human Genome Project is likely to be achieved by the systematic and continuous analysis of the resulting databases rather than by an individual use. The Coalition Proposal permitted research using existing databases unless the purpose of the researcher was direct competition with the database rights holder, an approach that fosters continuous discovery using databases.[63]

The provisions proposed in H.R. 354 and the Senate Discussion Draft also should be contrasted with the operation of fair use under the Copyright Act. Copyright fair use, like the H.R. 354 provision, depends on an after-the-fact balancing of factors to assess whether a subsequent otherwise infringing use is privileged. However, copyright law protects only the expression of a protected work and not the ideas or facts contained therein. Scientists may freely mine the world of existing copyright-protected works for the data upon which their research depends, without fear of liability. One of the significant aspects of a

---

[59] Section 1304(b)(2).

[60] See Reichman and Uhlir (1999), note 15, p. 815.

[61] Section 1403(b) and section 1303(a), respectively.

[62] Section 1403(a)(2).

[63] Section 1402(e).

database—a collection of facts—is that it can be difficult to draw the line between facts and expression. Neither H.R. 354 nor the Senate Discussion Draft purported to do so, but each prohibited the "extraction and use" of a substantial part (measured qualitatively or quantitatively) of a database.[64] The traditional methods of scientific research—as well as the mining of existing storehouses of ideas or facts (i.e., databases) upon which to build knowledge—would be placed at risk by these proposals. The Coalition Proposal, in addition to its narrow scope of initial prohibition, also expressly exempted from protection any "idea, fact, procedure, system, method of operation, concept, principle or discovery,"[65] consistent with the Copyright Act.[66] This breadth of exemption is also recommended by the committee.

Even the limited privileges offered to teachers and researchers in H.R. 354 and in the Senate Discussion Draft could be further undermined by the enforceability of contractual limitations on use of databases, or by technical measures that can prevent uses even if privileged. Neither H.R. 354 nor the Senate Discussion Draft imposed limits on the ability of a rights holder of a database to impose additional restrictions on use by contract or technical measures.[67] The Senate Discussion Draft did, however, attempt to temper unreasonable contractual overrides, particularly in cases of sole-source databases, by raising the issue of the potential application of the legal doctrine of misuse in the draft legislative history[68] and by including such issues as part of a required review of the effects of the legislation[69] (see the next section). The Coalition Proposal, on the other hand, adopted a "misuse" provision, which would authorize courts to deny relief to a database rights holder if "permitted acts" of database use are "frustrated by contractual arrangements or technological measures" or if "access to information necessary to research" is prevented.[70]

As noted above, research and education produce externalities that confer benefits on society at large. It is not clear that private parties through contracts will take these externalities into account when negotiating a license to use an existing database. Where this is true, licensors are likely to authorize fewer uses of their databases than would be socially optimal. Reliance on private decisions to ensure widespread availability of scientific and technical data runs the risk of interfering with research and education. Therefore, the creation of unprecedented new federal statutory rights for rights holders of databases must be balanced by

---

[64] H.R. 354, section 1402; Hatch Discussion Draft, section 1302.

[65] Section 1402(d).

[66] See 17 U.S.C., section 102(b).

[67] Section 1405(e) in H.R. 354, and section 1306(e) in the Hatch Discussion Draft.

[68] Section 1306 of the Proposed Conference Report language, pp. 36-37.

[69] See section 4(a), Study Regarding the Effect of the Act.

[70] Section 1407(b).

some affirmative duties not to unilaterally override by contract the public-interest exemptions and other permitted uses allowed under any new law.[71]

One approach would be to affirmatively state by legislative provision that traditional or customary scientific, educational, and research uses could not constitute infringements under unfair competition database legislation. This would hold true even if such traditional or customary public-interest uses caused potential or actual economic harm to the rights holder in a database. This approach closely parallels the existing "first sale" doctrine in that although current purchasing and subsequent lending of traditional books and journals (such as is done by libraries) may reduce the sales of these works and allegedly harm the economic interests of publishers, the sharing of legally acquired works is so important to society's scientific, educational, and social advancement that the potential or actual harms to authors and publishers are considered inconsequential when balanced against the benefits to society of allowing sharing. The customary and traditional practices of the research and educational communities were formed under the copyright law milieu, which achieved a careful balance of the rights of rights holders and users over time. The balance of interests struck by the law in paper publishing environments has worked well, and an analogous balance has to be developed in electronic sharing environments, particularly for scientific and technical databases.

It is important that any carved-out rights for traditional and customary scientific uses in any legislation that may be adopted not be able to be overridden or denied to scientists, educators, and other public-interest users through use of contracts.[72] In addition, integrative and derivative uses extending from other integrative and derivative uses have become one of the major methods of scientific inquiry today. As discussed in Chapter 1, scientific inquiry involves not only controlled observation and confirmation based on the published data, information, and knowledge from books, journals, and other intellectual works, but also the mining of electronic data sets that may have been gathered for scientific or other purposes. The sifting and winnowing of data and knowledge from all available sources contribute indispensably to the advancement of knowledge and the development of yet further derivative databases upon which others may build. Thus there is a need to minimize the barriers to access and use of facts and compilations of factual information, not increase them. It is extremely important

---

[71] For a discussion of proposed "public-interest unconscionability" clauses in licensing agreements for copyrighted works and noncopyrightable databases, see J.H. Reichman and Jonathan A. Franklin (1999), "Privately Legislated Intellectual Property Rights: Reconciling Freedom of Contract with Public Good Uses of Information," *U. Penn. L. Rev.*, Vol. 147, No. 4, pp. 929-951.

[72] For a discussion of legal and policy issues at the interface of contract law and copyright law see David Nimmer, Elliot Brown, and Gary N. Frischling (1999), "The Metamorphosis of Contract into Expand," *Cal. L. Rev.*, Vol. 87, p. 19, and Charles R. McManis (1999), "The Privatization (or "Shrink-Wrapping") of American Copyright Law," *Cal. L. Rev.*, Vol. 87, p. 1763.

that traditional and transformative uses of scientific and technical data be allowed by right without requiring permission of rights holders, similar to the legal situation for hard-copy documents under the "first sale" doctrine.

The "fair use" doctrine under copyright law and the "first sale" doctrine apply to all intellectual works and not just to "traditional or customary scientific, educational, and research uses" as proposed in the paragraphs above. Thus the balance between the exclusivity interests of the commercial community and the openness and sharing interests of the scientific and education communities is not fully preserved by the proposed legislation's exemptions for scientific inquiry.

The committee therefore recommends that any new legislation that may be adopted expressly continue to provide legal rights of access to and uses of proprietary databases equivalent to those that not-for-profit researchers, educators, and other public-interest users enjoyed under traditional or customary practice prior to enactment. Courts should be allowed to invalidate any non-bargained[73] licensing terms that are shown to interfere unduly with otherwise legislatively permitted customary uses by not-for-profit entities. Additional steps need to be taken by the government and by the research, education, and other public-interest communities, however, in the implementation of a new database protection regime to help ensure that the traditional and customary rights of public-interest data users are not unduly compromised. These steps are discussed in some detail in the section below titled "Assessment of Policy Options, with Recommendations for Government Action."

Periodic Assessments of Effects Under Any New Statute

As pointed out above in this chapter, the Commission of the European Communities (CEC) conducted no economic studies of the database industry, or of the potential effects of different models and provisions, to specifically support the drafting of the E.U. Database Directive. The only significant economic analysis done in the United States with regard to the pending legislation was an article commissioned by two of the principal supporters of H.R. 354, Reed-Elsevier, Inc., and the Thompson Corporation.[74] While neither the CEC nor the U.S. Congress has undertaken such studies in advance of their legislative initiatives, both legislative bodies have implicitly recognized that some negative effects are likely to be generated by this new legal regime. The E.U. Directive requires the CEC to submit to the European Parliament, the Council, and the Economic and Social Committee of the CEC a report

---

[73] By "non-bargained term" the committee means any term, usually contained in a standard form contract, over which, as a practical matter, no actual bargaining by the parties to the contract takes place.

[74] See Tyson and Sherry (1997), note 31.

. . . on the application of this Directive, in which, inter alia on the basis of specific information supplied by the Member States, it shall examine in particular the application of the sui generis right, including Articles 8 and 9, and shall verify especially whether the application of this right has led to abuse of dominant position or other interference with free competition which would justify appropriate measures being taken, including the establishment of non-voluntary licensing arrangements.[75]

Neither H.R. 354 nor its predecessor bills in the House, H.R. 2652 or Title V of H.R. 2281, had any provision for review of the economic effects of the bill on competition, consumers, or public-interest users. In contrast, the Senate Discussion Draft did provide for the conduct of a "Study Regarding the Effect of the Act" by the General Accounting Office, in consultation with the Register of Copyrights and the Department of Justice, within 5 years of enactment and every 10 years thereafter.[76] The issues for study that the Senate Discussion Draft would require are fully reproduced below, not only because they represent concerns regarding effects that might arise as a direct consequence of the enactment of this type of legislation, but also because they form the basis for a core set of questions that can be addressed independently by those studying the effects of the bill.

(b) ELEMENTS FOR CONSIDERATION—The study conducted under subsection (a) shall consider—

(1) The extent to which the ability of persons to engage in the permitted acts under this Act has been frustrated by contractual arrangements or technological measures;

(2) the extent to which information contained in databases that are the sole source of the information contained therein is made available through licensing or sale on reasonable terms and conditions;

(3) the extent to which the license or sale of information contained in databases protected under this Act has been conditioned on the acquisition or license of any other product or service, or on the performance of any action, not directly related to the license or sale;

(4) the extent to which the judicially-developed doctrines of misuse in other areas of the law have been extended to cases involving protection of databases under this Act;

---

[75] Article 16(3). Articles 8 and 9, which are referred to in Article 16(3), concern "Rights and obligations of lawful users" and "Exceptions to the sui generis right," respectively. It should be noted that until a very last-minute decision by the E.C. Council of Ministers, the proposed Database Directive contained a mandatory compulsory license for sole-source providers. See Reichman and Samuelson (1997), note 39, p. 87.

[76] Section 4(a).

(5) the extent, if any, to which the provisions of this Act constitute a barrier to entry, or have encouraged entry into, a relevant database market;

(6) the extent to which claims have been made that this Act prevented access to valuable information for research, competition, or innovation purposes and an evaluation of these claims;

(7) the extent to which enactment of this Act resulted in the creation of databases that otherwise would not exist; and

(8) such other matters necessary to accomplish the purpose of the report.[77]

This type of monitoring and review of the effects of database protection legislation should focus not only on national database activities, but on international ones as well, since the market for all online databases is inherently international, as are many S&T research activities. Although the committee believes that such periodic reviews would be important, particularly if they are not carried out prior to enactment of any new legislation, there are a number of other aspects to consider. There are several government entities that Congress might consider in addition to, or instead of, the three suggested above, including the Congressional Budget Office, the Department of Commerce, and the Federal Trade Commission, all of which would bring relevant expertise and interests to such a review. Also, because of the complexity of the issues to be examined, the rapidly changing nature of digital information technologies, and the lack of empirical data to fully support the analysis of these issues, the federal entities charged with doing the review should consider, in conjunction with the various stakeholder groups, what kind of data would be desirable to track and should initiate some means of doing so. Finally, the committee suggests that instead of an assessment of effects under the statute, Congress consider enacting a sunset provision that would take effect after a 5-year period and place the burden of proof and action on those who would want the legislation to continue. In light of the concerns about the possible unintended effects of such legislation, the rapid pace of change in digital information and network technologies, and the need to exercise due caution in the enactment of any such legislation, a sunset provision with the possibility of renewal may be the better option.

The committee therefore recommends that any new legislation provide either for a sunset provision with the possibility to renew, or for periodic assessments of the effects of new statutory database protection on competition in the database market and on consumers of databases, as well as on access to and use of data—including S&T data—by not-for-profit, public-interest users, in order to enable timely and appropriate revisions of legislation as needed.

---

[77] Id., (b).

Exemptions for Government Databases

As discussed throughout this report, S&T databases created either by the government or with government funding provide the largest and, in many scientific areas, the most important source of data for research and education. Moreover, existing U.S. law and policy prohibit proprietary protection of data or information created at the federal level, and generally limit such protection at the state and local levels as well. Maintaining these exemptions is of crucial importance, not only to public-interest users of databases in the research and educational communities but to all citizens. It therefore is not surprising that all three legislative proposals attempted to provide broad exemptions for government or government-funded databases.

Nevertheless, there are some notable differences among these proposals. Under H.R. 354, protection provided by the proposed legislation would not extend to collections of information gathered by or for a governmental entity, whether federal, state, or local.[78] For example, databases compiled by state or municipal governments in conjunction with their geographic information systems (GISs) would not be protected, although state and local governments would retain any copyright, contractual/licensing, trade secret, and technological protections that currently apply. H.R. 354 made it clear that the proposed legislation, and presumably federal copyright law, would preempt conflicting state laws.[79] This means that state misappropriation laws could not be applied by a state or local agency in a claim against a commercial business or vice versa.

H.R. 354 provided protection for collections of information created by state and federal educational institutions engaged in the course of education or scholarship.[80] Relative to civil law suits, the court would be required to reduce or remit entirely monetary relief "in any case in which a defendant believed and had reasonable grounds for believing that his or her conduct was permissible under this chapter, if the defendant was an employee or agent of a not-for-profit educational, scientific, or research institution, library, or archives acting within the scope of his or her employment."[81] A similar non-applicability provision would apply for criminal offenses under that proposed legislation.[82] This allowance, however, is merely an "innocent infringer" provision; once the not-for-profit researcher or educator is notified the first time, this immunity would be removed.

These liability relief provisions would not apply to state or local government agencies under H.R 354. Thus, if a state or municipal government GIS operation would use data from a commercial collection of information without permission,

---

[78] Section 1404(a)(1).
[79] Section 1405(b).
[80] Section 1404(a)(1).
[81] Section 1406(e).
[82] Section 1407(a)(2).

the government operation and employees acting even within the course of their employment would be subject to the liability provisions set forth in H.R. 354. Criminal violations would apply in situations where losses to the commercial company aggregated to more than $10,000 in a year or if the state or local government ran its operation for "direct or indirect commercial advantage or financial gain."[83] On the other hand, and again by example, commercial firms extracting government data from state or municipal GIS operations would not be subject to either civil or criminal liability provisions under that proposed legislation.

Under the Coalition Proposal, prohibitions against duplication would not apply to "government databases."[84] Here, however, "government database" was defined as being "a database (A) that has been collected or maintained by the United States of America; or (B) that is required by federal statute or regulation to be collected or maintained, to the extent so required."[85] Under the Coalition Proposal, state and local governments would gain most of the protections of the new database legislation, in a manner similar to that of commercial firms, and could choose to not avail themselves of the proposed legislative provisions at their option. This option is similar to the current situation in which many local governments choose not to seek copyright protection for their public records and databases. Under the liability provisions it is also clear that state and local governments would be in a position similar to that of commercial firms in being able to seek damages from private and government competitors who duplicate their data and compete with their income streams. Like H.R. 354, the Coalition Proposal stated that the new federal legislation would preempt conflicting state laws.

The research and education communities generally believe that the greatest benefits would accrue if state and local governments followed the federal government principles of "a strong freedom of information law, no government copyright, fees limited to recouping the cost of dissemination, and no restrictions on reuse."[86] This view appears to be gaining currency even in Europe. A green paper issued by the Commission of the European Communities in January 1999 comes to the conclusion that public-sector information is a key resource for Europe and suggests that E.U. nations should more closely follow the model of U.S. federal government policies with regard to promoting broader access to government databases.[87] Regardless of the merits of providing open access to

---

[83] Section 1407(a)(1).

[84] Section 1403(a).

[85] Section 1405(5).

[86] Peter N. Weiss and Peter Backlund (1997), "International Information Policy in Conflict: Open and Unrestricted Access versus Government Commercialization," in *Borders in Cyberspace: Information Policy and the Global Information Infrastructure*, Brian Kahin and Charles Nesson, eds., MIT Press, Cambridge, MA.

[87] DGXIII (1998), *Public Sector Information: A Key Resource for Europe*, "Green Paper on Public Sector Information in the Information Society," European Commission, Luxembourg. Available online at <www.echo.lu/legal/en/access.html>.

government information, the individual states in the United States traditionally have determined which policies they will follow in providing access to their state and local government information. The Coalition Proposal appeared to support continuation of this state and local government self-determination concept.

The Senate Discussion Draft more closely paralleled H.R. 354 than it did the Coalition Proposal in its effect on government data collections. For instance, the term "government databases" was again defined to include databases of government entities at all levels—federal, state, or local.[88] Therefore the protection and liability ramifications of the Senate Discussion Draft for governments would be similar in many respects to those of H.R. 354.

The Senate Discussion Draft theoretically would allow a library, archive, educational, scientific, or research institution to extract government data contained in a commercial database.[89] However, the requester would bear heavy burdens regarding data identification and proof of need, and the extraction would be allowed only in the unlikely event that the information was still available in its original format in the commercial database and separate from other portions of the commercial database. The requesting not-for-profit organization also would need to pay the costs of fulfilling the user request. From a practical perspective, it is difficult to envision the ability of libraries and educational institutions to successfully pursue such extractions.

The adoption of a strong standard of harm also could accelerate the privatization of government data dissemination with potentially negative results, as noted in Chapter 2. While privatization of some government functions may, under appropriate circumstances, produce net benefits, the committee urges caution in the context of government S&T database privatization in light of the public-good aspects of the data.[90] Because government databases are not and will not be legally protectable and the government's policy is to make public data broadly available, any entity can take the original data, add value, and redisseminate them as it wishes. In such cases, the original government data are nonetheless supposed to be (but are not always) maintained and to continue to be made available by the government source or archive.

There are some situations in which the government seeks to transfer the data dissemination function to a private-sector party, whether not-for-profit or commercial, on either a nonexclusive or an exclusive basis. The benefit of nonexclusive licensing is that several disseminators can compete in the market, and so tend toward providing access at a reasonable price for end users. In practice, however,

---

[88] Section 1301(6).

[89] Section 1305(b).

[90] See National Research Council (1997), *Bits of Power*, note 49, pp. 116-124 for a discussion of circumstances in which privatization of the government's data dissemination function is appropriate and inappropriate.

few organizations would be willing to enter into a formal agreement to handle a major data management and dissemination function on behalf of the government unless they could see some reasonable opportunity to at least recoup their operating expenses. Therefore, most such data management and dissemination functions are performed at university or other not-for-profit institution data centers that are partly subsidized by the originating government agency and are limited either by the terms of their agreement with the government or by their institutional charter (or both) to recovering operating costs, but prohibited from earning a profit.

In certain instances, government agencies have given exclusive licenses to commercial firms for dissemination of government data.[91] If the government agency subsequently does not continue to maintain and make available the original database, an exclusive license will result in a monopoly situation, which can lead to higher prices. Since stronger statutory database protection is likely to enhance the potential for profit by commercial data distributors, it also is likely to encourage the licensing of government data dissemination functions, perhaps on a de facto exclusive basis and without appropriate safeguards, thus defeating the existing open access and use law and policy of the U.S. government. As discussed in Chapter 2, the negative aspects of this trend are further exacerbated by other legislative initiatives that require government science agencies to purchase data from the private sector, rather than generate those same or similar data as a public good.

Finally, the exceptions for instructional and library uses articulated in the Senate Discussion Draft and in H.R. 354 were weak. Uses by academic institutions and libraries regarded as fair use under the Senate Discussion Draft are explicitly stated and represent a very narrow and qualified set of activities.[92] For example, display of the contents of a data collection during teaching in a normal class session that also is viewed by students at a distance enrolled in World Wide Web-based instruction would be in violation of the provisions of the Senate Discussion Draft. Numerous additional impositions on efficient approaches to instruction were raised by the Senate Discussion Draft; however, H.R. 354 did not even have a separate exemption for those activities.

The Senate Discussion Draft, and especially H.R. 354, focused primarily on strengthening proprietary protection without adequately balancing the public-interest, consumer, and commercial competitive values. The Coalition Proposal, by attempting to minimize social costs while providing some additional protection for commercial databases, appeared to arrive at a result that would be far more acceptable for maintaining the viability and vitality of the academic and

---

[91] See the Landsat privatization example discussed in National Research Council (1997), *Bits of Power*, note 49, pp. 121-123.

[92] Section 1307.

research communities upon which innovation and broad-based economic growth ultimately depend.

The committee thus recommends that although private-sector databases derived from government data should be eligible for protection, protection should not be extended to databases collected or maintained by the government. Any new legislation should expressly affirm the need for continuation of existing legal norms for wide distribution of government data and of data created pursuant to a government mandate or funding.

ASSESSMENT OF POLICY OPTIONS, WITH RECOMMENDATIONS FOR GOVERNMENT ACTION

In this section the committee discusses a number of actions that should be taken by various government institutions to help promote access to and use of S&T databases for the public interest. The areas addressed include promoting availability of government S&T data; maintaining nonexclusive rights in government-funded databases by not-for-profit institutions and their employees; organizing discussions of licensing terms for not-for-profit uses of commercial S&T databases; improving the understanding of complex economic aspects of S&T database activities; and promoting international access to S&T data. **Although the committee believes that its recommended actions in these areas ought to be undertaken whether or not any new statutory database protection is enacted by Congress, all of these actions will take on an increased urgency and importance if relatively strong new proprietary rights in databases are established by federal statute.**

Promoting Availability of Government Scientific and Technical Data

Increased proprietary protection for commercial databases could have a significant effect on government data collection and distribution efforts. Because researchers and educators likely would be more constrained in their use of data drawn from commercial databases, they might have to request additional funds for the purchase and administration of proprietary data or ask federal agencies to collect and maintain more S&T data on a nonproprietary basis. Thus budgetary strains could increase for federal agencies trying to meet the data needs of their own researchers as well as those related to fulfilling institutional mandates.

Under appropriate circumstances and conditions, government partnering with the private sector—especially not-for-profit institutions—in accomplishing data collection and maintenance can be highly beneficial and effective. The committee fully endorses the existing policy and practice of the federal government, as expressed through OMB Circular A-130, to make public S&T (and other) databases openly available at the lowest possible prices. It is through this policy of efficient-access pricing that the taxpayer derives maximum value from the government's very substantial investments in its collections of data.

Consistent with current practice, government S&T agencies must not make their databases—whether created and owned by them or under their control—available on an exclusive basis. They also should continue to maintain under their own control and archive all S&T databases that have value for research and that are otherwise being disseminated on behalf of the government by a private-sector organization or company. Such control should be maintained through physical possession or by appropriate contractual provisions. The long-term maintenance of public databases and archiving of data in readily accessible formats are essential to ensure their availability for reuse in future research or to confirm the results of research already conducted, among other uses.[93] Large quantities of government and government-funded data at all levels are lost, discarded, or rendered inaccessible owing to technological change or defects. Although this constitutes a major information management and policy issue in its own right that is beyond the scope of this report, the trend toward greater private-sector management and dissemination of public data—which the committee believes would increase under stronger statutory protection of databases—makes it even more important for government agencies to pay attention to this issue. Without adequate safeguards to ensure long-term preservation of public data created or disseminated on behalf of the government by private-sector entities, even larger amounts of such data may be lost or become inaccessible over time.

Finally, in making its data broadly available, the government should require that all private-sector disseminators or transformative users of its data identify the government source(s) of the data being used. Indeed, the same practice should be followed with regard to all sources of data from the private sector as well. Identifiers on privately disseminated government data will serve the objectives of providing notice to all users that they can contact the government agency source to obtain the original data, making the public aware of its government's activities, and giving proper credit where credit is due. Improving public awareness is an important objective, because all too often the public lacks a full appreciation of the benefits it derives from taxpayer-funded data-related activities.

The committee therefore recommends that the following actions be taken by all government entities. Scientific and technical data owned or controlled by the government should be made available for use by not-for-profit and commercial entities alike on a nonexclusive basis and should be disseminated to all users at no more than the marginal cost of reproduction and distribution, whenever possible. While the private sector's creation of derivative databases from government data should be encouraged, the source

---

[93] See generally National Research Council (1995), *Preserving Scientific Data on Our Physical Universe: A New Strategy for Archiving Our Nation's Scientific Resources*, National Academy Press, Washington, D.C.

of the original government data must ensure that those original data remain openly available. Any information product derived from a government database also should be required to carry an identifier stating the government source(s) used.

Maintaining Nonexclusive Rights by Not-for-Profits in Government-funded Databases

Science best advances by promoting a culture of openness and sharing, whereas individual commercial companies best advance by maintaining control and secrecy. The tension between these two cultures has been attenuated through the "first sale" doctrine, whereby a purchaser of a book, journal, or other intellectual work is free to use the facts and ideas in the work and the publisher is not able to prevent the purchaser from placing the intellectual work in a library or passing it on to another person for similar uses, regardless of the medium in which the work is presented.[94] As discussed in Chapter 3, this right is being superseded increasingly by licensing arrangements in the online digital environment. In the academic sector, such institutional licenses typically permit users to share with collaborators or colleagues on other campuses, so long as the sharing is not systematic. In fact, in some cases, universities are now able to negotiate licenses that "meet or exceed" their users' needs.[95] Nevertheless, in order to maintain a reasonable balance between the scientific and education communities' interests in openness and sharing, on the one hand, and the commercial community's interests in exclusivity on the other, some minimal constraints ought to be placed on the commercial community to guarantee researchers and educators access to and unfettered use of facts, data, and intellectual works published by their peers.

The existing practice in the publication of research results has been for researchers to pay page charges and to contractually give up exclusive copyright in their works in order to have their articles published in the primary journals read

---

[94] 17 U.S.C., section 109.

[95] Personal communication from Ann Okerson, Yale University, September 1999. An example of this type of licensing language is contained in Academic Press's IDEAL license (for 200+ journals):

> Copying and storing is limited to single copies of a reasonable number of individual items. Downloading an entire issue of a journal is not permitted. However, digital or print copies may be included in coursepacks and reserves, or in internal corporate training programs and drug application materials. Authorized Users may transmit downloaded copies of individual items to persons who are not Authorized Users for the purpose of scholarly communication, so long as the transmission is not done on a systematic basis.

At the same time, the traditional user rights under the first sale doctrine are in danger of being significantly eroded by the Uniform Computer Information Transactions Act, which is currently being considered for enactment at the state level. See generally, "A Guide to the Proposed Uniform Computer Information Transactions Act" at <www.2bguide.com>.

by their peers. Because authors transfer exclusive copyright in their work, they may be legally obligated to ask publishers for permission to distribute copies of their authored articles to their own students and to their close research associates. As one participant in the committee's January 1999 workshop wryly noted, "We have a great system: we pay to publish and we pay to get it back." In addition, the pricing structure that seems to maximize profits for commercial scientific publishers is one that limits acquisition of journals to the elite academic libraries and researchers that can afford them.[96]   Many academic institutions now have difficulty paying library subscription rates, and their researchers, professors, and students thus lack convenient access to many journals, even to those in which they publish.

Such concerns will only increase as electronic publishing becomes more widespread. It is already common practice in electronic publishing—and one of its tremendously productive features—to link electronic articles to the data sets upon which the research results depend. If legislation protecting databases is enacted, the current practice of requiring scientific authors to give up exclusive rights in their research articles on a take-it-or-leave-it basis could be extended to the data sets underlying the results reported in the research articles. Ceding control of databases created in the not-for-profit sector, especially those created with taxpayer support, to private-sector vendors that can establish their own terms for access to and use of the underlying research data is thus a major concern. In scientific disciplines where marketplace competition is highly con-strained or absent, there is a need to provide safeguards against monopolistic practices.

The committee therefore believes it important to initiate a "safety net" ap-proach in the digital database context to help preserve the balance previously provided by the "first sale" doctrine. This approach will help ensure public access to data and databases developed in whole or in substantial part at federal government expense. Databases developed primarily with government funds should not fall under the exclusive control of private parties such that dissemina-tion of the data to the public or other scientists is limited. Nor should public access to government-funded databases be highly constrained. The economic basis for funding science from governmental funds is that the research produces public goods. One researcher's use of these public goods does not decrease the value and benefits to others of the public goods.

Specifically, for any research accomplished wholly or in substantial part with federal funds, universities and not-for-profit organizations should be re-

---

[96] See Association for Research Libraries (1999), *ARL Statistics: 1997-98*, Martha Kyrillidou et al., eds., Association of Research Libraries, Washington, D.C., also available online at <www.arl.org/stats/arlstat/1997_t2.html>, showing trends in average rise in costs of serial (journal) subscriptions between 1986 and 1998, pp. 8-9. For a retrospective look at these issues see <www.lib.virginia.edu/mellon/mellon.html>.

quired by the funding agency to retain nonexclusive rights in any resulting databases. Under OMB Circular A-110, federal grant recipients have initial control over the intellectual property and databases that have been produced from their federally funded projects.[97] The primary concern under a new statutory regime is the inequality in bargaining power between large publishers and individual researchers and scientific authors. Based on past practices, the committee is concerned that many researchers may be required to give up exclusive rights in the databases produced at federal expense in return for having their research results published.

If new database legislation is enacted, publishers may request rights in both the intellectual work (i.e., typically the journal article), as well as rights to the collected data sets from which the intellectual work arose or upon which the work may depend and which might usefully be linked to the article in electronic publishing environments. If the negotiated contract provides for reasonable access to and use of the government-funded data for further scientific work, it is unlikely that the right of the researcher to independently provide access to the government-funded data would ever have to be invoked. However, the proposed provision provides a safety mechanism. The retained right of the researcher to distribute the data is likely to be invoked only in the unusual situation in which data gathered through a federally funded project or grant have been transferred through contract from an academic institution to a commercial entity with highly constrained access to and use of the data.

This approach has a relatively narrow application. The proposed requirement would not apply generally to copyrighted works that may have been produced using federal funds (e.g., research articles), nor would it apply to state or local government databases or to databases generally. The requirement also would not automatically apply to databases created with only partial (e.g., less than half) funding from the federal government, unless specifically agreed to by the parties. The provision would allow not-for-profit institutions and researchers to share underlying federally funded data with others regardless of contract provisions with the private sector, but would impose no affirmative requirement on them to share such data. Universities and other not-for-profits, of course, could not distribute any value-added features provided by the private-sector publisher to the government data unless agreed to by contract with that entity or otherwise permitted by law.

Based on the foregoing discussion, the committee recommends that federal funding agencies should require university and other not-for-profit researchers or their employing institutions that use federal funds, wholly or in

---

[97] Office of Management and Budget (1997), Circular A-110, "Uniform Administrative Requirements for Grants and Agreements with Institutions of Higher Education, Hospitals, and Other Not-for-profit Organizations," revised November 19, 1993; as further amended August 29, 1997.

substantial part, in creating databases not to grant exclusive rights to such databases when submitting them for publication or for incorporation into other databases.

Organizing Discussions of Licensing Terms for Not-for-Profit Uses of Commercial Scientific and Technical Databases

Whether or not new database protection legislation is adopted, the committee believes that representatives from the not-for-profit research and education communities should engage in a series of discussions with commercial database publishers and vendors in different market segments in order to achieve a better understanding of their respective needs and concerns and thus foster the development of mutually acceptable licensing terms that can reduce uncertainty and transaction costs. Such discussions would be especially important in the months and years immediately after enactment of a any new federal database statute, since there would be many definitions and concepts that would not have been fully defined and that would be subject to broadly divergent interpretations by different parties. One person's legitimate derivative use may be another's harmful infringement.

Previously established guidelines or understandings concerning copyrighted works will not in most cases be transferable to the database context and therefore will most likely confuse user communities, without the benefit of a fresh set of clarifying discussions. Further complications will arise if currently copyrighted works, such as journals, textbooks, reference books, and other anthologies, are also included in the definition of protected databases or "collections of information" under any new U.S. legislation, as they already are in the European Union. In addition to promoting some mutual understanding regarding licensing terms, clarifying discussions might help prevent unnecessary conflicts and litigation.

It is unrealistic to assume that a model contract or even standard individual contract terms could be developed to cover all or perhaps even most such transactions. As discussed in Chapter 1, a key characteristic of S&T data is the heterogeneity of data types, sources, and uses. The expectation of developing a one-size-fits-all approach would be not only illusory and impossible, but also ultimately harmful. To avoid becoming futile, discussions among stakeholders must be founded on realistic and well-focused objectives that would have a reasonable chance of success.

In establishing such discussions, it is essential that representatives of all major stakeholders be involved so that all relevant interests and viewpoints can be considered. For example, the committee would not endorse a process such as the one that resulted in the "Agreement on Guidelines for Classroom Copying."[98]

---

[98] For a history of academic fair use and classroom guidelines, see Kenneth D. Crews (1993), *Copyright, Fair Use and the Challenge of Universities*, University of Chicago Press, Chicago, IL.

That agreement has perhaps been workable for campus administrators, campus libraries, and the photocopying centers on campuses, but not for students and faculty, who were not involved as stakeholders in the discussions. Examples of both classroom guidelines[99] and of existing digital licensing terms and phrases, and their evaluation from the not-for-profit perspective,[100] may be found online on the World Wide Web already.

Private-sector S&T database producers and disseminators should remain cognizant of the social value of their products, particularly for not-for-profit research, education, and other public-interest uses. Database vendors whose primary source of revenue lies outside the not-for-profit S&T communities should endeavor to provide public-interest users access to their databases on favorable terms. Database vendors whose primary source of revenue is in the S&T research and education community should be encouraged to provide access on favorable terms once a reasonable return on investment has been achieved.

Indeed, as noted in Chapter 2, all pricing inhibits access, especially for those researchers who do not have adequate and strong institutional funding, whether academic, research institute, or industrial. Of course, the limitations are a matter of degree, depending on level and pattern of pricing. The goal should be to bring all sectors into a cooperative system in which data are made widely and readily available for scientific and educational use at as low a total cost (to the user population and society as a whole) as possible, and to do that within an environment that encourages, rather than inhibits, the inquisitiveness and inventiveness of the user while encouraging the entrepreneurship of suppliers. It is in the common interest of both database rights holders and users—and of society generally—to achieve a workable balance among the respective interests so that all legitimate rights remain reasonably protected.

The participants in these discussions would be primarily representatives of commercial S&T database disseminators and their government agency and not-for-profit-sector users. The committee makes its recommendation to the administration, rather than directly to those communities, however, because it believes that the discussions should be held under a convener such as the Copyright Office, which has the greatest subject matter expertise in these issues within the government. Such a focused venue would not only help stimulate progress on important issues, but would also mitigate the potential for accusations of collusion or conspiracy under federal antitrust laws. **The committee therefore recommends that the Copyright Office sponsor discussions between the representatives of private-sector producers of databases and user stakeholder representatives from government agencies and not-for-profit groups to help**

---

[99] Some examples of classroom guidelines may be found online at <fairuse.stanford.edu/library/index.html>.

[100] See <www.library.yale.edu>, and type "licensing" in the "Search" box.

develop a common understanding and optimal terms for the licensing of S&T databases and data products.

Improving the Understanding of Complex Economic Aspects of Scientific and Technical Database Activities

Although a detailed economic analysis is well beyond the scope of the committee's charge for this study, this report raises significant questions throughout regarding the adoption of statutory database protection; the economic underpinnings of different types and mixes of provisions, and the potential effects of both an overall statutory regime and specific provisions on various segments of the database industry and on the relationships among the different parties involved in creating, disseminating, and using S&T databases. Certainly, at a minimum, the questions raised in the E.U. Database Directive and in the Senate Discussion Draft, as well as any other questions that are ultimately identified in the course of the legislative process, should be the subject of more detailed study in advance of any legislatively mandated report on effects of increased protection. Such a study would help provide a comprehensive base of knowledge with which to officially evaluate the effects of new statutory protection of databases. In addition to these broad economic issues, the committee suggests that research be devoted to, among others, the following specific issues affecting the creation, dissemination, and use of S&T databases by the government and by the not-for-profit and for-profit sectors:

1. Is there an increase in the number of databases available that can be related to (caused or affected by) an increase in database protection?

2. Is there an increase in the costs (or a reduction in the amount) of scientific research that can be related to an increase in database protection?

3. What is the role of licensing restrictions on access to and use of databases for research and education activities and what are the effects?

4. What are the effects of sole-source databases on S&T niche markets and on the level of scientific research?

5. What are the trends in privatization of government S&T data, and what effects has privatization had on access to and use of such data?

6. What have been the effects of the E.U. Database Directive on access to and use of European S&T data by the European research and education community, and by the U.S. government and by the U.S. research and education community?

The committee recommends that the Congressional Research Service, the National Science Foundation, the Department of Commerce, and other federal science agencies, as considered appropriate, should undertake and fund external research that investigates the changing and complex economic

aspects of S&T database activities, particularly in the context of any new legislative database protection measures that may be enacted and in support of the legislative principle recommended above regarding the conduct of periodic assessments of the effects of any new statutory protection of databases.

Promoting International Access to Scientific and Technical Data

It is a well-known truism that science knows no boundaries and that practically all research that is conducted on an open basis also involves international collaboration to some degree. Some research, such as that in observational space and Earth sciences, is inherently international and cannot be conducted successfully without either the collection of global data or access to foreign databases.[101] As a result, the U.S. government science agencies in recent decades have concluded thousands of bilateral and multilateral general S&T cooperation and specific research program agreements.[102] These agreements will take on added significance with the implementation of the E.U. Database Directive and with possible adoption of restrictive database protection legislation in the United States and elsewhere, since the negotiated terms of those agreements can specify the terms under which databases related to the research in question can be accessed and used. As the world's largest producer and disseminator of S&T data, the U.S. government has significant leverage in negotiating appropriate terms for the exchange and use of public data with other nations.

At the same time, the committee agrees with the Administration's concerns regarding the E.U. Directive's reciprocity provision and supports the U.S. Trade Representative's (USTR's) placement of that topic on the Administration's 1998 Special 301 Review.[103] The committee would go one step further, however, and suggest that the USTR and other appropriate entities within the Administration negotiate with the Commission of the European Communities to review and revise its E.U. Directive, based on the substantial criticisms of that new legal regime in this report and in other position statements and articles cited above. If the U.S. Congress enacts a new database protection statute based on properly balanced unfair competition principles, the committee urges the USTR, the U.S. Patent and Trademark Office, and other appropriate administration officials to

---

[101] For example, for a comprehensive listing of most internationally available data sets from space missions, see the NASA Goddard Space Flight Center's National Space Science Data Center home page online at <nssdc.gsfc.nasa.gov>. For a listing of many international Web sites covering all aspects of Earth science data, see the NASA Global Change Master Directory online at <http://gcmd.gsfc.nasa.gov/pointers/pointwais.html>.

[102] For a discussion of some of the large international research programs, see National Research Council (1997), *Bits of Power*, note 49, pp. 58-61.

[103] See Pincus statement, March 18, 1999, Hearing, note 7, p. 33.

promote that statute as a model for international database protection within the World Intellectual Property Organization.

The committee recommends that all departments and agencies of the federal government should continue to adopt international S&T agreements that include provisions to facilitate access to S&T data across national boundaries and should conduct periodic reviews of international policies and agreements to promote conformity to the above principles.

In addition, the committee recommends that the U.S. government should negotiate with the Commission of the European Communities to revise its highly protectionist E.U. Database Directive.

RECOMMENDED APPROACH FOR THE NOT-FOR-PROFIT SCIENTIFIC AND TECHNICAL COMMUNITY

Finally, there is the question of what the research and education community should do in the event that highly restrictive statutory protection of databases is enacted by Congress. Certainly, leaders in all the major not-for-profit research, higher education, and library associations and in many individual institutions have voiced their concerns about the legislative proposals that have been introduced in the Committee on the Judiciary in the House of Representatives in both the 105th and 106th Congress.[104] As the various critics and this report point out, such a new statutory regime could have many negative effects, among them significant changes in the terms for access to and use of databases sold or licensed by the commercial sector, the possibility of increased economic exploitation on a proprietary basis of heretofore openly available S&T databases by not-for-profit researchers and educators and their institutions, and the stimulation of further privatization of such public-good activities by the government.

The question thus raised is what actions the not-for-profit community should take on its own behalf if restrictive new provisions are enacted that encourage or exacerbate these negative effects. Proponents of new legislation rightly point out that any new law would not *require* individuals or organizations to make use of the new protections, and that if the not-for-profits are concerned about these legal developments as a matter of principle, they should resist the temptation to adopt proprietary restrictions on their databases. The committee agrees that much of the responsibility for maintaining a policy of full and open data availability in the academic community rests with the community, which will have to act to ensure continuation of the broad sharing of data and research results. Nevertheless, the

---

[104] See the testimony given by Wulf, Reichman, and Neal at the October 23, 1997, Hearing, note 8; by Stewart at the February 12, 1998, Hearing, note 9; and by Lederberg, Phelps, and Neal at the March 18, 1999, Hearing, note 29. See also the position statement signed by representatives of many of these organizations, note 7.

committee also believes that the pressures to commercialize and privatize currently open data sources would increase inevitably under a regime such as the E.U. Directive or the one proposed by H.R. 354, and that fully maintaining the customary or traditional approaches to data exchange would prove to be difficult. In the event that the universe of public domain S&T data is found to be shrinking unacceptably, additional defensive measures may have to be taken to reinvigorate a robust public-interest sector for such data.

Therefore, as its last recommendation, the committee urges that the not-for-profit S&T community continue to promote and adhere to the policy of full and open exchange of data at both the national and international levels.

Appendixes

Appendix A

Biographical Sketches of Committee Members

STUDY COMMITTEE

ROBERT J. SERAFIN (chair) is the director of the National Center for Atmospheric Research (NCAR). Dr. Serafin began his career at Hazeltine Research Corporation where he worked on the design and development of high-resolution radar systems. This was followed by 10 years at the IIT Research Institute and Illinois Institute of Technology. He then joined NCAR as manager of the Field Observing Facility in 1973 and in 1980 became director of the Atmospheric Technology Division, which is responsible for all of NCAR's observational research and research support facilities, used by scientists in universities and laboratories throughout the world. In 1989 he was appointed as NCAR's director. The holder of three patents, Dr. Serafin has published approximately 50 technical and scientific papers and established the *Journal of Atmospheric and Oceanic Technology*, and was its co-editor for several years. He has served on several National Research Council (NRC) panels and committees, and he chaired the NRC committee on National Weather Service Modernization. He is a member of the National Academy of Engineering, a fellow of the American Meteorological Society, and a senior member of the Institute of Electrical and Electronics Engineers. Dr. Serafin received his BS, MS, and PhD degrees in electrical engineering from Notre Dame University, Northwestern University, and Illinois Institute of Technology, respectively.

I. TROTTER HARDY is professor of law at the College of William & Mary School of Law, where he specializes in intellectual property law, law and computers, and tort law. He recently wrote a major report, "Sketching the Future of Copyright in a Networked World," for the Copyright Office and has published extensively on issues relating to intellectual property law in the digital environment. Mr. Hardy holds a BA from the University of Virginia, an MS from American University, and a JD from Duke University, Order of the Coif.

MAUREEN C. KELLY is vice president for planning at BIOSIS, the largest abstracting and indexing service for the life sciences community. She has worked in different capacities for BIOSIS since 1969. Previously she had production responsibility for the bibliographic and scientific content of BIOSIS products. While in that position, she led the team that developed the system for capturing and managing indexing data in support of BIOSIS's new relational indexing. Ms. Kelly has authored a number of papers on managing and accessing biological information. She is currently secretary of the American Association for the Advancement of Science Section on Information, Computing, and Communication. She has served on various professional society research and publishing committees, including participating in the National Academy of Sciences E-Journal Summit meetings over the past two years. Ms. Kelly has a BA degree from Rutgers University.

PETER R. LEAVITT is a consultant and former chairman and chief executive officer of Weather Services Corp., where he has developed online real-time meteorological databases for national and international agricultural and commodity services. He has a BS in meteorology from the Massachusetts Institute of Technology. Mr. Leavitt served previously on two NRC committees, as well as on several government advisory committees regarding data use and research issues in meteorology.

LEE E. LIMBIRD is associate vice chancellor for research at Vanderbilt University and chair of the Department of Pharmacology. Her responsibilities as associate vice chancellor include development of new intra- and interinstitutional initiatives for research, with a focus on research development in genetics and genomics; neuroscience; and structural biology, broadly defined to include biophysics and bioengineering. The Office of Grants Management and Technology Transfer also represents areas of her responsibility. Dr. Limbird received a BA in chemistry from the College of Wooster and a PhD in biochemistry from the University of North Carolina. Her area of research has been in the molecular pathways of signal transduction by G Protein-coupled receptors using biochemical, cellular, and genetic strategies, including genetically modified mice.

PHILIP LOFTUS is vice president and director of Worldwide Information Services Architecture and Technology for Glaxo Wellcome where he is responsible for both the information services infrastructure and global information management. From 1996 to 1998, he served as vice president and director of Worldwide R&D Information Systems and was responsible for developing and implementing a global information system strategy for R&D. Prior to that, he was executive director for Research Information Systems at Merck Research Laboratories, and from 1976 to 1993 he was a vice president for R&D Information Systems and a computational scientist at ICI. Dr. Loftus has a BSc in chemistry and a PhD in conformational isomerism from the University of Liverpool, and he was a Fullbright Hayes Postdoctoral Fellow at the California Institute of Technology in 1974-1975. He holds a postgraduate certificate in education from the University of Liverpool. He has published extensively in the area of information technology applications for pharmaceutical research.

HARLAN J. ONSRUD is professor in the Department of Spatial Information Science and Engineering at the University of Maine and a research scientist with the National Center for Geographic Information and Analysis (NCGIA). He received BS and MS degrees in civil engineering from the University of Wisconsin and a JD from the University of Wisconsin Law School. His research focuses on (1) analysis of legal and institutional issues affecting the creation and use of digital databases and the sharing of geographic information, (2) assessing utilization of GIS and the social impacts of the technology, and (3) developing and assessing strategies for supporting the diffusion of geographic information innovations. Mr. Onsrud has co-led major multiyear NCGIA research initiatives on the use and value of geographic information, institutions sharing geographic information, and law, information policy, and spatial databases. Mr. Onsrud is a licensed engineer, lawyer, and land surveyor.

HARVEY S. PERLMAN is a professor of law and former dean of the University of Nebraska College of Law. He is an expert in trademark law and unfair competition law. In addition to writing many articles in these areas, Mr. Perlman has co-authored *Legal Regulation of the Competitive Process: Cases, Materials and Notes on Unfair Business Practices*, which is now in its sixth edition under the title *Intellectual Property and Unfair Competition* (1998). He also was the co-reporter for the American Law Institute's *Restatement (Third) of Unfair Competition* and is a member of the National Conference of Commissioners on Uniform State Laws, which has been considering changes to the Uniform Commercial Code Article 2(B) regarding private contracts for intellectual property. Mr. Perlman received his BA and JD from the University of Nebraska in 1963 and 1966, respectively.

ROBERTA P. SAXON is a patent agent at Skjerven, Morrill, MacPherson, Franklin & Friel, LLP, a law firm specializing in intellectual property in San Jose, California. Prior to that, she was director of the chemistry laboratory at SRI International, where she supervised research in advanced materials, atmospheric chemistry, computational chemistry, and atomic, molecular, and optical physics and performed research in those areas for more than 20 years. Dr. Saxon has a BA in chemistry from Cornell University and an MS and a PhD in chemical physics from the University of Chicago. She is vice chair of the Panel on Public Affairs for the American Physical Society, and she previously served on an NRC study for a research strategy for atomic, molecular, and optical sciences.

SUZANNE SCOTCHMER is a professor of economics and public policy at the University of California, Berkeley. Her broad fields of research are in economic theory and industrial organization, with current emphasis on intellectual property, particularly as it relates to cumulative innovations, digital content, and decentralized mechanisms by which firms share information. Dr. Scotchmer received her PhD in economics from the University of California, Berkeley in 1980 and her MA in statistics in 1979.

MARK STEFIK is a principal scientist at the Xerox Palo Alto Research Center, where he focuses on trusted system approaches for creating, protecting, and reusing digital property in the network context. His current and past research activities include research on reasoning with constraints, and paradigms of programming, as well as applications of artificial intelligence and computer science to problems in molecular genetics, VLSI circuit design, configuration of computer systems, and systems for supporting collaborative processes in work groups. Dr. Stefik's book, *Internet Dreams: Archetypes, Myths, and Metaphors,* was published by MIT Press in 1996. Dr. Stefik received his BS and PhD from Stanford University.

MARTHA E. WILLIAMS is director of the Information Retrieval Research Lab and a professor of Information Science at the University of Illinois at Urbana-Champaign. Her research interests include digital database management, online retrieval systems, systems analysis and design, chemical information systems, and electronic publishing. She has published widely on these topics and has been editor of the *Annual Review of Information Sciences and Technology* (since 1975), *Computer Readable Databases: A Directory & Data Sourcebook* (1976-1987), and *Online Review* (since 1977). Professor Williams was chair of the Board of Engineering Information, Inc., from 1980 to 1988, was appointed to the National Library of Medicine's Board of Regents from 1978 to 1981 and served as chair of the board in 1981. In addition, she served on several NRC committees, including the Numerical Data Advisory Board (1979-1982). She has an AB from Barat College and an MA from Loyola University.

STUDY DIRECTOR

PAUL F. UHLIR is director of international scientific and technical information programs at the U.S. National Academy of Sciences/National Research Council (National Academies) in Washington, D.C., where he directs science and technology policy studies for the federal government. His current area of emphasis is issues at the interface of science, technology, and law, with primary focus on scientific data and information policy, and on the relationship of intellectual property law to R&D policy. Mr. Uhlir is also director of the U.S. National Committee for CODATA. From 1991 to 1998, he was associate executive director of the Commission on Physical Sciences, Mathematics, and Applications, and from 1985 to 1991 he was senior program officer at the Space Studies Board, where he worked on solar system exploration and Earth remote sensing studies for NASA. Before joining the National Academies, he was a foreign affairs officer at the National Oceanic and Atmospheric Administration in the Department of Commerce, where he worked on meteorological and land remote sensing law and policy issues. He is the author or editor of more than 50 books, reports, and articles. Mr. Uhlir has a BA in history from the University of Oregon, and a JD and an MA in international relations from the University of San Diego.

Appendix B

Workshop Agenda and Participants

WORKSHOP AGENDA

Thursday, January 14

8:00 am **Continental breakfast**

8:30 **A. Introductory remarks**
Robert Serafin, Study Chair

8:45 ***Keynote Address***
Q. Todd Dickinson
Commissioner of Patents and Trademarks (Acting), Department of
Commerce

B. Summary of S&T databases to be discussed at the workshop

9:00 **Geographic Data Panel**
Moderator: Harlan Onsrud, Associate Professor, University of Maine
* *Government-sector data activity:* Barbara Ryan, Associate
Director for Operations, U.S. Geological Survey, Department
of the Interior

- *Not-for-profit sector data activity:* James Brunt, Associate Director for Information Management, Long-Term Ecological Research Network Office, University of New Mexico[1]
- *Commercial-sector data activity:* Barry Glick, former President and CEO, GeoSystems Global Corp.

9:45   **Genomic Data Panel**
Moderator: Philip Loftus, Vice President and Director, Glaxo Wellcome

- *Government-sector data activity:* James Ostell, Chief, Information Engineering Branch, National Center for Biotechnology Information, National Library of Medicine, National Institutes of Health
- *Not-for-profit-sector data activity:* Chris Overton, Director, Center for Bioinformatics, University of Pennsylvania
- *Commercial-sector data activity:* Myra Williams, President and CEO, Molecular Applications Group

10:30   **Break**

10:45   **Chemical and Chemical Engineering Data Panel**
Moderator: Roberta Saxon, Patent Agent, Skjerven, Morrill, MacPherson, et al.

- *Government-sector data activity:* Richard Kayser, Chief, Physical and Chemical Properties Division, National Institute of Standards and Technology, Department of Commerce
- *Not-for-profit-sector data activity:* James Lohr, Director, Information Industry Relations, Chemical Abstracts Service, American Chemical Society
- *Commercial-sector data activity:* Leslie Singer, President, ISI, Inc.

11:30   **Meteorological Data Panel**
Moderator: Robert Serafin, Director, National Center for Atmospheric Research

- *Government-sector data activity:* Ken Hadeen, Director (retired), National Climatic Data Center, National Oceanic and Atmospheric Administration, Department of Commerce
- *Not-for-profit-sector data activity:* David Fulker, Director, Unidata Program, University Corporation for Atmospheric Research

---

[1] Dr. Brunt was unable to attend the workshop due to illness.

• *Commercial-sector data activity:* Robert Brammer, Vice President and Chief Technology Officer, TASC

12:15 pm **Lunch**

1:15 **C. Economic factors in production/dissemination/use of S&T databases in the public and private sectors**
Moderator: Suzanne Scotchmer, Professor, UC Berkeley
Speaker: Richard Gilbert, Professor, UC Berkeley

2:15 **D. Overview of technologies for protecting and misappropriating digital IPR: the current situation and future prospects**
Moderator: Mark Stefik, Principal Scientist, Xerox PARC
Speaker: Teresa Lunt, Principal Scientist, Xerox PARC (by video)

2:45 **Break**

3:00 **E.1 Summary overview of existing and proposed IPR regimes for databases**
• **The status quo**
• *Sui generis* **property rights model**
• **Unfair competition/misappropriation model**
Moderator: Harvey Perlman, Professor, College of Law, University of Nebraska
Speaker: Marybeth Peters, Register of Copyrights, Library of Congress

3:45 **E.2 Summary of federal government information law and data policies**
Speaker: Justin Hughes, Attorney, Patent and Trademark Office, Department of Commerce

4:00 **F. Breakout sessions on the existing legal and technical situation**

4:15 **Individual breakout sessions**

1) Government-sector data panel
Moderator: Shelton Alexander, Professor, Pennsylvania State University
Rapporteur: Suzanne Scotchmer, Professor, UC Berkeley
Panelists: • *Barbara Ryan*, Associate Director of Operations, U.S. Geological Survey

- *James Ostell*, Chief, Information Engineering Branch, National Center for Biotechnology Information, NLM/NIH
- *Richard Kayser*, Chief, Physical and Chemical Properties Division, National Institute of Standards and Technology
- *Kenneth Hadeen*, Director (retired), National Climatic Data Center

2) Not-for-profit-sector data panel

Moderator: Maureen Kelly, Vice President for Planning, BIOSIS
Rapporteur: Jerome Reichman, Professor, Vanderbilt University School of Law
Panelists:
- *James Brunt*, Associate Director for Information Management, Long-Term Ecological Research Network Office, University of New Mexico[2]
- *Chris Overton*, Director, Center for Bioinformatics, University of Pennsylvania
- *James Lohr*, Director, Information Industry Relations, Chemical Abstracts Service, American Chemical Society
- *David Fulker*, Director, Unidata Program, UCAR

3) Commercial-sector data panel

Moderator: Robert Serafin, Director, National Center for Atmospheric Research
Rapporteur: Mark Stefik, Principal Scientist, Xerox PARC
Panelists:
- *Barry Glick*, former President and CEO, GeoSystems Global Corp.
- *Myra Williams*, President and CEO, Molecular Applications Group
- *Leslie Singer*, President, ISI, Inc.
- *Robert Brammer*, Vice President and Chief Technology Officer, TASC

5:45 **Adjourn**

5:45 -
6:45 pm **Reception**

---

[2] Dr. Brunt was unable to attend the workshop due to illness.

Friday, January 15

8:00 am **Continental breakfast**

8:30   **G. Summary reports by rapporteurs from previous day's breakouts**

9:20   **H. Instructions by workshop chair and move to breakout rooms**

9:30   **I. Breakout sessions**

Session 1:   Congress decides to enact a strong property rights model protecting databases

Moderator: Paul Uhlir, Study Director, National Research Council
Rapporteur: Peter Leavitt, Consultant
Panelists: • *Ken Hadeen*, Director (retired), National Climatic Data Center
             • *David Fulker*, Director, Unidata Program, UCAR
             • *Robert Brammer*, Vice President and Chief Technology Officer, TASC
             • *Jon Baumgarten*, Attorney, Proskauer Rose LLP
             • *Peter Jaszi*, Professor, American University School of Law
             • *James Neal*, Director, Johns Hopkins University Library
             • *Ferris Webster*, Professor, University of Delaware

Session 2:   Congress decides to enact an unfair competition model protecting databases

Moderator: Harvey Perlman, Professor, College of Law, University of Nebraska
Rapporteur: Philip Loftus, Vice President and Director, Glaxo Wellcome
Panelists: • *Dennis Benson,* Chief Information Resources Branch, National Center for Biotechnology Information, NLM/NIH
             • *Chris Overton*, Director, Center for Bioinformatics, University of Pennsylvania
             • *Myra Williams*, President and CEO, Molecular Applications Group
             • *Michael Klipper*, Attorney, Meyer & Klipper, PLLC
             • *Jonathan Band*, Attorney, Morrison & Foerster, LLP

- *Thomas Rindfleisch*, Director, Medical Library, Stanford University

Session 3:   Promoting access to and use of government S&T data for the public interest—an assessment of legal and policy options

Moderator: Harlan Onsrud, Associate Professor, University of Maine

Rapporteur: Shelton Alexander, Professor, Pennsylvania State University

Panelists:
- *Barbara Ryan*, Associate Director for Operations, U.S. Geological Survey[3]
- *James Brunt*, Associate Director for Information Management, Long-Term Ecological Research Network Office, University of New Mexico[4]
- *Barry Glick*, former President and CEO, GeoSystems Global Corp.
- *Peter Weiss,* Senior Policy Analyst, Office of Management and Budget[5]
- *Prue Adler,* Assistant Executive Director, Federal Relations and Information Policy, Association of Research Libraries
- *Eric Massant*, Director of Government and Industry Affairs, Reed Elsevier, Inc.
- *Tim Foresman*, Director, Spatial Analysis Lab, University of Maryland
- *Kenneth Frazier*, Director, University of Wisconsin Libraries

Session 4:   Promoting access to and use of not-for-profit-sector S&T data for the public interest—an assessment of legal and policy options

Moderator: Martha Williams, Professor and Director, Information Retrieval Research Lab, University of Illinois at Urbana-Champaign

Rapporteur: Roberta Saxon, Patent Agent, Skjerven, Morrill, MacPherson et al.

Panelists:
- *Richard Kayser*, Chief, Physical and Chemical Properties Division, NIST

---

[3] Ms. Ryan was unable to attend this session due to inclement weather.

[4] Dr. Brunt was unable to attend the workshop due to illness.

[5] Mr. Weiss was unable to attend this session due to inclement weather.

- *James Lohr*, Director, Information Industry Relations, Chemical Abstracts Service, American Chemical Society
- *Leslie Singer*, President, ISI, Inc.
- *Allan Adler*, Vice President for Governmental and Legislative Affairs, Association of American Publishers, Inc.[6]
- *Jerome Reichman*, Professor, Vanderbilt University School of Law
- *R. Stephen Berry*, Professor, University of Chicago

10:45   **Break**

11:00   **Breakout session discussions (continued)**

12:45 pm **Lunch**

1:45   **J. Rapporteurs' summary of breakout panel results**

2:45   **Discussion of results with workshop participants**

3:45   **K. Concluding remarks**
        Robert Serafin, Chair

4:00 pm **End of public workshop**

WORKSHOP PARTICIPANTS

Allan Adler, Association of American Publishers, Inc.
Prue Adler, Association of Research Libraries
Shelton Alexander, Pennsylvania State University
Dave Applegate (affiliation unknown)
Christopher Ashley, National Science Foundation
Mary Baish, American Association of Law
Jonathan Band, Morrison & Foerster, LLP
Winona Barker, National Biomedical Research Foundation
Ed Barron, Senate Committee on the Judiciary
Barbara Bauldock, U.S. Geological Survey
Jon Baumgarten, Proskauer Rose, LLP

---

[6] Mr. Adler was unable to attend this session due to inclement weather.

Dennis Benson, National Library of Medicine
R. Stephen Berry, University of Chicago
Robert Brammer, TASC
Lisa Brooks, National Institutes of Health
Francis Buckley, Jr., U.S. Government Printing Office
Mark Burnham, California Institute of Technology
Bonnie Carroll, Information International Associates, Inc.
William Cohen, Federal Trade Commission
Kathy Covert, Federal Geographic Data Committee
Karen Dacres, National Oceanic and Atmospheric Administration
Judge Edward Damich, U.S. Court of Federal Claims
Matthew Davis (affiliation unknown)
Deveny Deck, Vanderbilt University
Paul DeGiusti, Information Industry Association
Q. Todd Dickinson, U.S. Patent and Trademark Office
Anita Eisenstadt, National Science Foundation
Adam Eisgrau, American Library Association
Julie Esanu, National Research Council
Bob Etkins, National Oceanic and Atmospheric Administration
Eric Fischer, Library of Congress
Peter Folger, American Geophysical Union
Tim Foresman, University of Maryland
Mark Frankel, American Association for the Advancement of Science
Kenneth Frazier, University of Wisconsin Libraries
David Fulker, University Corporation for Atmospheric Research
Carole Ganz-Brown, National Science Foundation
Richard Gilbert, University of California, Berkeley
Paul Gilman, Celera Genomics, Inc.
Barry Glick, Consultant
Kenneth Hadeen, National Climatic Data Center (retired)
Kelley Heilman, Maryland State Department of Health
Stephen Heinig, Association of American Medical Colleges
Stephen Heller, National Institute of Standards and Technology
Mike Hoffman (affiliation unknown)
Justin Hughes, U.S. Patent and Trademark Office
Peter Jaszi, American University School of Law
Brian Kahin, Office of Science and Technology Policy
Richard Kayser, National Institute of Standards and Technology
Chris Kelly, U.S. Department of Justice
Maureen Kelly, BIOSIS
Michael Keplinger, U.S. Patent and Trademark Office
Ehsan Khan, U.S. Department of Energy
Michael Klipper, Meyer & Klipper, PLLC

Makoto Kono, Fujitsu, Ltd.
Stephen Koslow, National Institute of Mental Health
Patrice Laget, Delegation of the European Commission
Richard Lambert, National Institutes of Health
Charles Larson, Industrial Research Institute, Inc.
Peter Leavitt, Consultant
Robert Ledley, Georgetown University
Lynn Levine, Warren Publishing
David Lide, Consultant
Anne Linn, National Research Council
Joan Lippincott, Coalition for Networked Information
Philip Loftus, Glaxo Wellcome
James Lohr, Chemical Abstracts Service
Joe Martinez, U.S. Department of Energy
Eric Massant, Reed Elsevier, Inc.
Stephen Maurer, Attorney
Gilles McDougall, Industry Canada
Bruce McDowell, National Academy of Public Administration
Shelia McGarr (affiliation unknown)
Theodore Miles, National Science Foundation
John Moeller, U.S. Geological Survey
Christopher Mohr, Meyer & Klipper, PLLC
Kurt Molholm, Defense Technical Information Center
James Neal, Johns Hopkins University Library
Judge Pauline Newman, U.S. Court of Appeals for the Federal Circuit
Goetz Oertel, Association of Universities for Research in Astronomy
Harlan Onsrud, University of Maine
James Ostell, National Institutes of Health
G. Christian Overton, University of Pennsylvania
Bob Palmer, U.S. House of Representatives
Harvey Perlman, University of Nebraska
Shira Perlmutter, U.S. Copyright Office
Marybeth Peters, U.S. Copyright Office
Larry Pettinger, U.S. Geological Survey
Tony Reichardt, *Nature*
Jerome Reichman, Vanderbilt University School of Law
Thomas Rindfleisch, Stanford University
Hedy Rossmeissl, U.S. Geological Survey
John Rumble, National Institute of Standards and Technology
Barbara Ryan, U.S. Geological Survey
Carolina Saez, U.S. Copyright Office
Roberta Saxon, Skjerven, Morrill, MacPherson, Franklin & Friel, LLP
Terri Scanlan, National Research Council

Jean Schiro-Zavela, National Oceanic and Atmospheric Administration
Harold Schoolman, National Library of Medicine
Suzanne Scotchmer, University of California, Berkeley
Robert Serafin, National Center for Atmospheric Research
Leslie Singer, Institute for Scientific Information, Inc.
Mark Smith, American Association of University Professors
Mark Stefik, Xerox Palo Alto Research Center
Charles Sturrock, National Institute of Standards and Technology
Ambassador James Sweeney, Consultant
Margaret Thomson, U.S. Department of Energy
Paul Uhlir, National Research Council
John Vaughn, Association of American Universities
Ferris Webster, University of Delaware
Peter Weiss, Office of Management and Budget
Pamela Whitney, National Research Council
Martha Williams, University of Illinois at Urbana-Champaign
Myra Williams, Molecular Applications Group
James Wilson, House Committee on Science
Richard Witmer, U.S. Geological Survey
Barbara Wright, National Research Council
Susan Zevin, National Oceanic and Atmospheric Administration

APPENDIX C

Workshop *Proceedings*— Listing of Contents

Part I—Workshop Presentations

1   Introductory Remarks
    Robert Serafin

2   Keynote Address
    Q. Todd Dickinson

3   Characteristics of Scientific and Technical Databases
    Geographic Data Panel
    Genomic Data Panel
    Chemical and Chemical Engineering Data Panel
    Meteorological Data Panel

4   Economic Factors in the Production, Dissemination, and
        Use of Scientific and Technical Databases
    Richard Gilbert

NOTE: For the full text of the committee's *Proceedings*, see National Research Council (1999), *Proceedings of the Workshop on Promoting Access to Scientific and Technical Data for the Public Interest: An Assessment of Policy Options,* National Academy Press, Washington, D.C., available only online at < http://www.nap.edu>.

Appendix D

European Union Directive on the Legal Protection of Databases

DIRECTIVE 96/9/EC OF THE EUROPEAN PARLIAMENT AND OF THE COUNCIL
of 11 March 1996 on the legal protection of databases

THE EUROPEAN PARLIAMENT AND THE COUNCIL OF THE EUROPEAN
UNION,
Having regard to the Treaty establishing the European Community, and in particular
Article 57 (2), 66 and 100a thereof,
Having regard to the proposal from the Commission (1),
Having regard to the opinion of the Economic and Social Committee (2),
Acting in accordance with the procedure laid down in Article 189b of the Treaty (3),
(1) Whereas databases are at present not sufficiently protected in all Member States
by existing legislation; whereas such protection, where it exists, has different
attributes;
(2) Whereas such differences in the legal protection of databases offered by the
legislation of the Member States have direct negative effects on the functioning of the
internal market as regards databases and in particular on the freedom of natural and
legal persons to provide on-line database goods and services on the basis of
harmonized legal arrangements throughout the Community; whereas such differences

NOTE: An official version of this document can be found online at the EUR-LEX Web site at
<http://europa.eu.int/eur-lex/en/lif/dat/1996/en_396L0009.html>.

The material presented in this appendix has been reprinted from electronic files available on the
Internet and is intended for use as a general reference, and not for legal research or other work
requiring authenticated primary sources.

could well become more pronounced as Member States introduce new legislation in this field, which is now taking on an increasingly international dimension;

(3) Whereas existing differences distorting the functioning of the internal market need to be removed and new ones prevented from arising, while differences not adversely affecting the functioning of the internal market or the development of an information market within the Community need not be removed or prevented from arising;

(4) Whereas copyright protection for databases exists in varying forms in the Member States according to legislation or case-law, and whereas, if differences in legislation in the scope and conditions of protection remain between the Member States, such unharmonized intellectual property rights can have the effect of preventing the free movement of goods or services within the Community;

(5) Whereas copyright remains an appropriate form of exclusive right for authors who have created databases;

(6) Whereas, nevertheless, in the absence of a harmonized system of unfair-competition legislation or of case-law, other measures are required in addition to prevent the unauthorized extraction and/or re-utilization of the contents of a database;

(7) Whereas the making of databases requires the investment of considerable human, technical and financial resources while such databases can be copied or accessed at a fraction of the cost needed to design them independently;

(8) Whereas the unauthorized extraction and/or re-utilization of the contents of a database constitute acts which can have serious economic and technical consequences;

(9) Whereas databases are a vital tool in the development of an information market within the Community; whereas this tool will also be of use in many other fields;

(10) Whereas the exponential growth, in the Community and worldwide, in the amount of information generated and processed annually in all sectors of commerce and industry calls for investment in all the Member States in advanced information processing systems;

(11) Whereas there is at present a very great imbalance in the level of investment in the database sector both as between the Member States and between the Community and the world's largest database-producing third countries;

(12) Whereas such an investment in modern information storage and processing systems will not take place within the Community unless a stable and uniform legal protection regime is introduced for the protection of the rights of makers of databases;

(13) Whereas this Directive protects collections, sometimes called 'compilations', of works, data or other materials which are arranged, stored and accessed by means which include electronic, electromagnetic or electro-optical processes or analogous processes;

(14) Whereas protection under this Directive should be extended to cover non-electronic databases;

(15) Whereas the criteria used to determine whether a database should be protected by copyright should be defined to the fact that the selection or the arrangement of the contents of the database is the author's own intellectual creation; whereas such protection should cover the structure of the database;

(16) Whereas no criterion other than originality in the sense of the author's intellectual creation should be applied to determine the eligibility of the database for copyright protection, and in particular no aesthetic or qualitative criteria should be applied;

(17) Whereas the term 'database' should be understood to include literary, artistic, musical or other collections of works or collections of other material such as texts, sound, images, numbers, facts, and data; whereas it should cover collections of independent works, data or other materials which are systematically or methodically arranged and can be individually accessed; whereas this means that a recording or an audiovisual, cinematographic, literary or musical work as such does not fall within the scope of this Directive;

(18) Whereas this Directive is without prejudice to the freedom of authors to decide whether, or in what manner, they will allow their works to be included in a database, in particular whether or not the authorization given is exclusive; whereas the protection of databases by the sui generis right is without prejudice to existing rights over their contents, and whereas in particular where an author or the holder of a related right permits some of his works or subject matter to be included in a database pursuant to a non-exclusive agreement, a third party may make use of those works or subject matter subject to the required consent of the author or of the holder of the related right without the sui generis right of the maker of the database being invoked to prevent him doing so, on condition that those works or subject matter are neither extracted from the database nor re-utilized on the basis thereof;

(19) Whereas, as a rule, the compilation of several recordings of musical performances on a CD does not come within the scope of this Directive, both because, as a compilation, it does not meet the conditions for copyright protection and because it does not represent a substantial enough investment to be eligible under the sui generis right;

(20) Whereas protection under this Directive may also apply to the materials necessary for the operation or consultation of certain databases such as thesaurus and indexation systems;

(21) Whereas the protection provided for in this Directive relates to databases in which works, data or other materials have been arranged systematically or methodically; whereas it is not necessary for those materials to have been physically stored in an organized manner;

(22) Whereas electronic databases within the meaning of this Directive may also include devices such as CD-ROM and CD-i;

(23) Whereas the term 'database' should not be taken to extend to computer programs used in the making or operation of a database, which are protected by Council Directive 91/250/EEC of 14 May 1991 on the legal protection of computer programs (4);

(24) Whereas the rental and lending of databases in the field of copyright and related rights are governed exclusively by Council Directive 92/100/EEC of 19 November 1992 on rental right and lending right and on certain rights related to copyright in the field of intellectual property (5);

(25) Whereas the term of copyright is already governed by Council Directive 93/98/EEC of 29 October 1993 harmonizing the term of protection of copyright and certain related rights (6);

(26) Whereas works protected by copyright and subject matter protected by related rights, which are incorporated into a database, remain nevertheless protected by the respective exclusive rights and may not be incorporated into, or extracted from, the database without the permission of the rightholder or his successors in title;

(27) Whereas copyright in such works and related rights in subject matter thus incorporated into a database are in no way affected by the existence of a separate right in the selection or arrangement of these works and subject matter in a database;

(28) Whereas the moral rights of the natural person who created the database belong to the author and should be exercised according to the legislation of the Member States and the provisions of the Berne Convention for the Protection of Literary and Artistic Works; whereas such moral rights remain outside the scope of this Directive;

(29) Whereas the arrangements applicable to databases created by employees are left to the discretion of the Member States; whereas, therefore nothing in this Directive prevents Member States from stipulating in their legislation that where a database is created by an employee in the execution of his duties or following the instructions given by his employer, the employer exclusively shall be entitled to exercise all economic rights in the database so created, unless otherwise provided by contract;

(30) Whereas the author's exclusive rights should include the right to determine the way in which his work is exploited and by whom, and in particular to control the distribution of his work to unauthorized persons;

(31) Whereas the copyright protection of databases includes making databases available by means other than the distribution of copies;

(32) Whereas Member States are required to ensure that their national provisions are at least materially equivalent in the case of such acts subject to restrictions as are provided for by this Directive;

(33) Whereas the question of exhaustion of the right of distribution does not arise in the case of on-line databases, which come within the field of provision of services; whereas this also applies with regard to a material copy of such a database made by the user of such a service with the consent of the rightholder; whereas, unlike CD-ROM or CD-i, where the intellectual property is incorporated in a material medium, namely an item of goods, every on-line service is in fact an act which will have to be subject to authorization where the copyright so provides;

(34) Whereas, nevertheless, once the rightholder has chosen to make available a copy of the database to a user, whether by an on-line service or by other means of distribution, that lawful user must be able to access and use the database for the purposes and in the way set out in the agreement with the rightholder, even if such access and use necessitate performance of otherwise restricted acts;

(35) Whereas a list should be drawn up of exceptions to restricted acts, taking into account the fact that copyright as covered by this Directive applies only to the selection or arrangements of the contents of a database; whereas Member States should be given the option of providing for such exceptions in certain cases; whereas, however, this option should be exercised in accordance with the Berne Convention and to the extent that the exceptions relate to the structure of the database; whereas a distinction should be drawn between exceptions for private use and exceptions for reproduction for private purposes, which concerns provisions under national legislation of some Member States on levies on blank media or recording equipment;

(36) Whereas the term 'scientific research' within the meaning of this Directive covers both the natural sciences and the human sciences;

(37) Whereas Article 10 (1) of the Berne Convention is not affected by this Directive;

(38) Whereas the increasing use of digital recording technology exposes the database maker to the risk that the contents of his database may be copied and rearranged electronically, without his authorization, to produce a database of identical content which, however, does not infringe any copyright in the arrangement of his database;

(39) Whereas, in addition to aiming to protect the copyright in the original selection or arrangement of the contents of a database, this Directive seeks to safeguard the position of makers of databases against misappropriation of the results of the financial and professional investment made in obtaining and collection [*sic*] the contents by protecting the whole or substantial parts of a database against certain acts by a user or competitor;

(40) Whereas the object of this sui generis right is to ensure protection of any investment in obtaining, verifying or presenting the contents of a database for the limited duration of the right; whereas such investment may consist in the deployment of financial resources and/or the expending of time, effort and energy;

(41) Whereas the objective of the sui generis right is to give the maker of a database the option of preventing the unauthorized extraction and/or re-utilization of all or a substantial part of the contents of that database; whereas the maker of a database is the person who takes the initiative and the risk of investing; whereas this excludes subcontractors in particular from the definition of maker;

(42) Whereas the special right to prevent unauthorized extraction and/or re-utilization relates to acts by the user which go beyond his legitimate rights and thereby harm the investment; whereas the right to prohibit extraction and/or re-utilization of all or a substantial part of the contents relates not only to the manufacture of a parasitical competing product but also to any user who, through his acts, causes significant detriment, evaluated qualitatively or quantitatively, to the investment;

(43) Whereas, in the case of on-line transmission, the right to prohibit re-utilization is not exhausted either as regards the database or as regards a material copy of the database or of part thereof made by the addressee of the transmission with the consent of the rightholder;

(44) Whereas, when on-screen display of the contents of a database necessitates the permanent or temporary transfer of all or a substantial part of such contents to another medium, that act should be subject to authorization by the rightholder;

(45) Whereas the right to prevent unauthorized extraction and/or re-utilization does not in any way constitute an extension of copyright protection to mere facts or data;

(46) Whereas the existence of a right to prevent the unauthorized extraction and/or re-utilization of the whole or a substantial part of works, data or materials from a database should not give rise to the creation of a new right in the works, data or materials themselves;

(47) Whereas, in the interests of competition between suppliers of information products and services, protection by the sui generis right must not be afforded in such a way as to facilitate abuses of a dominant position, in particular as regards the creation and distribution of new products and services which have an intellectual,

documentary, technical, economic or commercial added value; whereas, therefore, the provisions of this Directive are without prejudice to the application of Community or national competition rules;

(48) Whereas the objective of this Directive, which is to afford an appropriate and uniform level of protection of databases as a means to secure the remuneration of the maker of the database, is different from the aim of Directive 95/46/EC of the European Parliament and of the Council of 24 October 1995 on the protection of individuals with regard to the processing of personal data and on the free movement of such data (7), which is to guarantee free circulation of personal data on the basis of harmonized rules designed to protect fundamental rights, notably the right to privacy which is recognized in Article 8 of the European Convention for the Protection of Human Rights and Fundamental Freedoms; whereas the provisions of this Directive are without prejudice to data protection legislation;

(49) Whereas, notwithstanding the right to prevent extraction and/or re-utilization of all or a substantial part of a database, it should be laid down that the maker of a database or rightholder may not prevent a lawful user of the database from extracting and re-utilizing insubstantial parts; whereas, however, that user may not unreasonably prejudice either the legitimate interests of the holder of the sui generis right or the holder of copyright or a related right in respect of the works or subject matter contained in the database;

(50) Whereas the Member States should be given the option of providing for exceptions to the right to prevent the unauthorized extraction and/or re-utilization of a substantial part of the contents of a database in the case of extraction for private purposes, for the purposes of illustration for teaching or scientific research, or where extraction and/or re-utilization are/is carried out in the interests of public security or for the purposes of an administrative or judicial procedure; whereas such operations must not prejudice the exclusive rights of the maker to exploit the database and their purpose must not be commercial;

(51) Whereas the Member States, where they avail themselves of the option to permit a lawful user of a database to extract a substantial part of the contents for the purposes of illustration for teaching or scientific research, may limit that permission to certain categories of teaching or scientific research institution;

(52) Whereas those Member States which have specific rules providing for a right comparable to the sui generis right provided for in this Directive should be permitted to retain, as far as the new right is concerned, the exceptions traditionally specified by such rules;

(53) Whereas the burden of proof regarding the date of completion of the making of a database lies with the maker of the database;

(54) Whereas the burden of proof that the criteria exist for concluding that a substantial modification of the contents of a database is to be regarded as a substantial new investment lies with the maker of the database resulting from such investment;

(55) Whereas a substantial new investment involving a new term of protection may include a substantial verification of the contents of the database;

(56) Whereas the right to prevent unauthorized extraction and/or re-utilization in respect of a database should apply to databases whose makers are nationals or habitual residents of third countries or to those produced by legal persons not

established in a Member State, within the meaning of the Treaty, only if such third countries offer comparable protection to databases produced by nationals of a Member State or persons who have their habitual residence in the territory of the Community;

(57) Whereas, in addition to remedies provided under the legislation of the Member States for infringements of copyright or other rights, Member States should provide for appropriate remedies against unauthorized extraction and/or re-utilization of the contents of a database;

(58) Whereas, in addition to the protection given under this Directive to the structure of the database by copyright, and to its contents against unauthorized extraction and/or re-utilization under the sui generis right, other legal provisions in the Member States relevant to the supply of database goods and services continue to apply;

(59) Whereas this Directive is without prejudice to the application to databases composed of audiovisual works of any rules recognized by a Member State's legislation concerning the broadcasting of audiovisual programmes;

(60) Whereas some Member States currently protect under copyright arrangements databases which do not meet the criteria for eligibility for copyright protection laid down in this Directive; whereas, even if the databases concerned are eligible for protection under the right laid down in this Directive to prevent unauthorized extraction and/or re-utilization of their contents, the term of protection under that right is considerably shorter than that which they enjoy under the national arrangements currently in force; whereas harmonization of the criteria for determining whether a database is to be protected by copyright may not have the effect of reducing the term of protection currently enjoyed by the rightholders concerned; whereas a derogation should be laid down to that effect; whereas the effects of such derogation must be confined to the territories of the Member States concerned,

HAVE ADOPTED THIS DIRECTIVE:

CHAPTER I

SCOPE

Article 1

Scope
1. This Directive concerns the legal protection of databases in any form.
2. For the purposes of this Directive, 'database' shall mean a collection of independent works, data or other materials arranged in a systematic or methodical way and individually accessible by electronic or other means.
3. Protection under this Directive shall not apply to computer programs used in the making or operation of databases accessible by electronic means.

Article 2

Limitations on the scope
This Directive shall apply without prejudice to Community provisions relating to:

(a) the legal protection of computer programs;

(b) rental right, lending right and certain rights related to copyright in the field of intellectual property;

(c) the term of protection of copyright and certain related rights.

CHAPTER II

COPYRIGHT

Article 3

Object of protection

1. In accordance with this Directive, databases which, by reason of the selection or arrangement of their contents, constitute the author's own intellectual creation shall be protected as such by copyright. No other criteria shall be applied to determine their eligibility for that protection.

2. The copyright protection of databases provided for by this Directive shall not extend to their contents and shall be without prejudice to any rights subsisting in those contents themselves.

Article 4

Database authorship

1. The author of a database shall be the natural person or group of natural persons who created the base or, where the legislation of the Member States so permits, the legal person designated as the rightholder by that legislation.

2. Where collective works are recognized by the legislation of a Member State, the economic rights shall be owned by the person holding the copyright.

3. In respect of a database created by a group of natural persons jointly, the exclusive rights shall be owned jointly.

Article 5

Restricted acts

In respect of the expression of the database which is protectable by copyright, the author of a database shall have the exclusive right to carry out or to authorize:

(a) temporary or permanent reproduction by any means and in any form, in whole or in part;

(b) translation, adaptation, arrangement and any other alteration;

(c) any form of distribution to the public of the database or of copies thereof. The first sale in the Community of a copy of the database by the rightholder or with his consent shall exhaust the right to control resale of that copy within the Community;

(d) any communication, display or performance to the public;

(e) any reproduction, distribution, communication, display or performance to the public of the results of the acts referred to in (b).

Article 6

Exceptions to restricted acts
1. The performance by the lawful user of a database or of a copy thereof of any of the acts listed in Article 5 which is necessary for the purposes of access to the contents of the databases and normal use of the contents by the lawful user shall not require the authorization of the author of the database. Where the lawful user is authorized to use only part of the database, this provision shall apply only to that part.
2. Member States shall have the option of providing for limitations on the rights set out in Article 5 in the following cases:
(a) in the case of reproduction for private purposes of a non-electronic database;
(b) where there is use for the sole purpose of illustration for teaching or scientific research, as long as the source is indicated and to the extent justified by the non-commercial purpose to be achieved;
(c) where there is use for the purposes of public security of [*sic*] for the purposes of an administrative or judicial procedure;
(d) where other exceptions to copyright which are traditionally authorized under national law are involved, without prejudice to points (a), (b) and (c).
3. In accordance with the Berne Convention for the protection of Literary and Artistic Works, this Article may not be interpreted in such a way as to allow its application to be used in a manner which unreasonably prejudices the rightholder's legitimate interests or conflicts with normal exploitation of the database.

CHAPTER III

SUI GENERIS RIGHT

Article 7

Object of protection
1. Member States shall provide for a right for the maker of a database which shows that there has been qualitatively and/or quantitatively a substantial investment in either the obtaining, verification or presentation of the contents to prevent extraction and/or re-utilization of the whole or of a substantial part, evaluated qualitatively and/or quantitatively, of the contents of that database.
2. For the purposes of this Chapter:
(a) 'extraction' shall mean the permanent or temporary transfer of all or a substantial part of the contents of a database to another medium by any means or in any form;
(b) 're-utilization' shall mean any form of making available to the public all or a substantial part of the contents of a database by the distribution of copies, by renting, by on-line or other forms of transmission. The first sale of a copy of a database within the Community by the rightholder or with his consent shall exhaust the right to control resale of that copy within the Community;
Public lending is not an act of extraction or re-utilization.
3. The right referred to in paragraph 1 may be transferred, assigned or granted under contractual licence.

4. The right provided for in paragraph 1 shall apply irrespective of the eligibility of that database for protection by copyright or by other rights. Moreover, it shall apply irrespective of eligibility of the contents of that database for protection by copyright or by other rights. Protection of databases under the right provided for in paragraph 1 shall be without prejudice to rights existing in respect of their contents.

5. The repeated and systematic extraction and/or re-utilization of insubstantial parts of the contents of the database implying acts which conflict with a normal exploitation of that database or which unreasonably prejudice the legitimate interests of the maker of the database shall not be permitted.

Article 8

Rights and obligations of lawful users

1. The maker of a database which is made available to the public in whatever manner may not prevent a lawful user of the database from extracting and/or re-utilizing insubstantial parts of its contents, evaluated qualitatively and/or quantitatively, for any purposes whatsoever. Where the lawful user is authorized to extract and/or re-utilize only part of the database, this paragraph shall apply only to that part.

2. A lawful user of a database which is made available to the public in whatever manner may not perform acts which conflict with normal exploitation of the database or unreasonably prejudice the legitimate interests of the maker of the database.

3. A lawful user of a database which is made available to the public in any manner may not cause prejudice to the holder of a copyright or related right in respect of the works or subject matter contained in the database.

Article 9

Exceptions to the sui generis right

Member States may stipulate that lawful users of a database which is made available to the public in whatever manner may, without the authorization of its maker, extract or re-utilize a substantial part of its contents:

(a) in the case of extraction for private purposes of the contents of a non-electronic database;

(b) in the case of extraction for the purposes of illustration for teaching or scientific research, as long as the source is indicated and to the extent justified by the non-commercial purpose to be achieved;

(c) in the case of extraction and/or re-utilization for the purposes of public security or an administrative or judicial procedure.

Article 10

Term of protection

1. The right provided for in Article 7 shall run from the date of completion of the making of the database. It shall expire fifteen years from the first of January of the year following the date of completion.

2. In the case of a database which is made available to the public in whatever manner before expiry of the period provided for in paragraph 1, the term of protection by

that right shall expire fifteen years from the first of January of the year following the date when the database was first made available to the public.

3. Any substantial change, evaluated qualitatively or quantitatively, to the contents of a database, including any substantial change resulting from the accumulation of successive additions, deletions or alterations, which would result in the database being considered to be a substantial new investment, evaluated qualitatively or quantitatively, shall qualify the database resulting from that investment for its own term of protection.

Article 11

Beneficiaries of protection under the sui generis right
1. The right provided for in Article 7 shall apply to database [*sic*] whose makers or rightholders are nationals of a Member State or who have their habitual residence in the territory of the Community.
2. Paragraph 1 shall also apply to companies and firms formed in accordance with the law of a Member State and having their registered office, central administration or principal place of business within the Community; however, where such a company or firm has only its registered office in the territory of the Community, its operations must be genuinely linked on an ongoing basis with the economy of a Member State.
3. Agreements extending the right provided for in Article 7 to databases made in third countries and falling outside the provisions of paragraphs 1 and 2 shall be concluded by the Council acting on a proposal from the Commission. The term of any protection extended to databases by virtue of that procedure shall not exceed that available pursuant to Article 10.

CHAPTER IV

COMMON PROVISIONS

Article 12

Remedies
Member States shall provide appropriate remedies in respect of infringements of the rights provided for in this Directive.

Article 13

Continued application of other legal provisions
This Directive shall be without prejudice to provisions concerning in particular copyright, rights related to copyright or any other rights or obligations subsisting in the data, works or other materials incorporated into a database, patent rights, trade marks, design rights, the protection of national treasures, laws on restrictive practices and unfair competition, trade secrets, security, confidentiality, data protection and privacy, access to public documents, and the law of contract.

Article 14

Application over time
1. Protection pursuant to this Directive as regards copyright shall also be available in respect of databases created prior to the date referred to Article 16 (1) which on that date fulfil the requirements laid down in this Directive as regards copyright protection of databases.
2. Notwithstanding paragraph 1, where a database protected under copyright arrangements in a Member State on the date of publication of this Directive does not fulfil the eligibility criteria for copyright protection laid down in Article 3 (1), this Directive shall not result in any curtailing in that Member State of the remaining term of protection afforded under those arrangements.
3. Protection pursuant to the provisions of this Directive as regards the right provided for in Article 7 shall also be available in respect of databases the making of which was completed not more than fifteen years prior to the date referred to in Article 16 (1) and which on that date fulfil the requirements laid down in Article 7.
4. The protection provided for in paragraphs 1 and 3 shall be without prejudice to any acts concluded and rights acquired before the date referred to in those paragraphs.
5. In the case of a database the making of which was completed not more than fifteen years prior to the date referred to in Article 16 (1), the term of protection by the right provided for in Article 7 shall expire fifteen years from the first of January following that date.

Article 15

Binding nature of certain provisions
Any contractual provision contrary to Articles 6 (1) and 8 shall be null and void.

Article 16

Final provisions
1. Member States shall bring into force the laws, regulations and administrative provisions necessary to comply with this Directive before 1 January 1998.
When Member States adopt these provisions, they shall contain a reference to this Directive or shall be accompanied by such reference on the occasion of their official publication. The methods of making such reference shall be laid down by Member States.
2. Member States shall communicate to the Commission the text of the provisions of domestic law which they adopt in the field governed by this Directive.
3. Not later than at the end of the third year after the date referred to in paragraph 1, and every three years thereafter, the Commission shall submit to the European Parliament, the Council and the Economic and Social Committee a report on the application of this Directive, in which, inter alia, on the basis of specific information supplied by the Member States, it shall examine in particular the application of the sui generis right, including Articles 8 and 9, and shall verify especially whether the

application of this right has led to abuse of a dominant position or other interference with free competition which would justify appropriate measures being taken, including the establishment of non-voluntary licensing arrangements. Where necessary, it shall submit proposals for adjustment of this Directive in line with developments in the area of databases.

Article 17
This Directive is addressed to the Member States.

Done at Strasbourg, 11 March 1996.
For the European Parliament
The President
K. HÄNSCH
For the Council
The President
L. DINI

(1) OJ No C 156, 23. 6. 1992, p. 4 and
OJ No C 308, 15. 11. 1993, p. 1.
(2) OJ No C 19, 25. 1. 1993, p. 3.
(3) Opinion of the European Parliament of 23 June 1993 (OJ No C 194, 19. 7. 1993, p. 144), Common Position of the Council of 10 July 1995 (OJ No C 288, 30. 10. 1995, p. 14), Decision of the European Parliament of 14 December 1995 (OJ No C 17, 22 1. 1996) and Council Decision of 26 February 1996.
(4) OJ No L 122, 17. 5. 1991, p. 42. Directive as last amended by Directive 93/98/EEC (OJ No L 290, 24. 11. 1993, p. 9.)
(5) OJ No L 346, 27. 11. 1992, p. 61.
(6) OJ No L 290, 24. 11. 1993, p. 9.
(7) OJ No L 281, 23. 11. 1995, p. 31.